T0210542

SpringerBriefs in Computer Science

More information about this series at http://www.springer.com/series/10028

Dan Wang • Zhu Han

Sublinear Algorithms
for Big Data Applications

 Springer

Dan Wang
Department of Computing
The Hong Kong Polytechnic University
Kowloon, Hong Kong, SAR

Zhu Han
Department of Engineering
University of Houston
Houston, TX, USA

ISSN 2191-5768   ISSN 2191-5776   (electronic)
SpringerBriefs in Computer Science
ISBN 978-3-319-20447-5   ISBN 978-3-319-20448-2   (eBook)
DOI 10.1007/978-3-319-20448-2

Library of Congress Control Number: 2015943617

Springer Cham Heidelberg New York Dordrecht London

Printed on acid-free paper

Springer International Publishing AG Switzerland is part of Springer Science+Business Media (www.springer.com)

Dedicate to my family, Dan Wang
Dedicate to my family, Zhu Han

Preface

In recent years, we see a tremendously increasing amount of data. A fundamental challenge is how these data can be processed efficiently and effectively. On one hand, many applications are looking for solid foundations; and on the other hand, many theories may find new meanings. In this book, we study one specific advancement in theoretical computer science, the sublinear algorithms and how they can be used to solve big data application problems. Sublinear algorithms, as what the name shows, solve problems using less than linear time or space as against to the input size, with provable theoretical bounds. Sublinear algorithms were initially derived from approximation algorithms in the context of randomization. While the spirit of sublinear algorithms fit for big data application, the research of sublinear algorithms is often restricted within theoretical computer sciences. Wide application of sublinear algorithms, especially in the form of current big data applications, is still in its infancy. In this book, we take a step towards bridging such gap. We first present the foundation of sublinear algorithms. This includes the key ingredients and the common techniques for deriving the sublinear algorithm bounds. We then present how to apply sublinear algorithms to three big data application domains, namely, wireless sensor networks, big data processing in MapReduce, and smart grids. We show how problems are formalized, solved, and evaluated, such that the research results of sublinear algorithms from the theoretical computer sciences can be linked with real-world problems.

We would like to thank Prof. Sherman Shen for his great help in publishing this book. This book is also supported by US NSF CMMI-1434789, CNS-1443917, ECCS-1405121, CNS-1265268, and CNS- 0953377, National Natural Science Foundation of China (No. 61272464), and RGC/GRF PolyU 5264/13E.

Kowloon, Hong Kong   Dan Wang
Houston, TX, USA   Zhu Han

Contents

Chapter 1
Introduction

1.1 Big Data: The New Frontier

In February 2010, National Centers for Disease Control and Prevention (CDC) identified an outbreak of flu in the mid-Atlantic regions of the United States. However, 2 weeks earlier, Google Flu Trends [1] had already predicted such an outbreak. By no means does Google have more expertise in the medical domain than the CDC. However, Google was able to predict the outbreak early because it uses big data analytics. Google establishes an association between outbreaks of flu and user queries, e.g., on throat pain, fever, and so on. The association is then used to predict the flu outbreak events. Intuitively, an association means that if event A (e.g., a certain combination of queries) happens, event B (e.g., a flu outbreak) will happen (e.g., with high probability). One important feature of such analytics is that the association can only be established when the data is big. When the data is small, such as a combination of a few user queries, it may not expose any connection with a flu outbreak. Google applied millions of models to the huge number of queries that it has. The aforementioned prediction of flue by Google is an early example of the power of big data analytics, and the impact of which has been profound.

The number of successful big data applications is increasing. For example, Amazon uses massive historical shipment tracking data to recommend goods to targeted customers. Indeed such "Target Marketing" has been adopted and is being carried out by all business sectors that have penetrated all aspects of our life. We see personalized recommendations from the web pages we commonly visit, from the social network applications we use daily, and from the online game stores we frequently access. In smart cities, data on people, the environment, and the operational components of the city are collected and analyzed (see Fig. 1.1). More specifically, data on traffic and air quality reports are used to determine the causes of heavy air pollution [3], and the huge amount of data on bird migration paths are analyzed to predict H5N1 bird flu [4]. In the area of B2B, there are startup companies (e.g., MoleMart, MolBase) that analyze huge amount

© The Author(s) 2015
D. Wang, Z. Han, *Sublinear Algorithms for Big Data Applications*,
SpringerBriefs in Computer Science, DOI 10.1007/978-3-319-20448-2_1

Fig. 1.1 Smart City, a big vision of the future where people, environment, and city operational components are in harmony. One key to achieve this is big data analytics, where data of people, environment and city operational components are collected and analyzed. The data variety is diverse, the volume is big, the collection velocity can be high, and the veracity may be problematic; yet handling these appropriately, the value can be significant

of data on pharmaceutical, biological, and chemical related industries. Accurate connections between buyers and vendors are established and the risk to companies of overstocking or understocking is reduced. This has lead to cost reductions of more than ten times compared to current B2B intermediaries.

The expectations for the future are even greater. Today, scientists, engineers, educators, citizens, and decision-makers have unprecedented amounts and types of data available to them. Data come from many disparate sources, including scientific instruments, medical devices, telescopes, microscopes, and satellites; digital media including text, video, audio, email, weblogs, twitter feeds, image collections, click streams, and financial transactions; dynamic sensors, social, and other types of networks; scientific simulations, models, and surveys; or from computational analysis of observational data. Data can be temporal, spatial, or dynamic; and structured or unstructured. Information and knowledge derived from data can differ in representation, complexity, granularity, context, provenance, reliability, trustworthiness, and scope. Data can also differ in the rate at which they are generated and accessed.

On the one hand, the enriched data provide opportunities for new observations, new associations, and new correlations to be made, which leading to added value and new business opportunities. On the other hand, big data poses a fundamental challenge to the efficient processing of data. Gigabytes, terabytes or even petabytes of data need to be processed. People commonly refer to the *volume, velocity, variety, veracity* and *value* of data as the 5-V model. Again, take the smart city as an example (see Fig. 1.1). The big vision of the future is that the people, environment, and city operational components of the city be in harmony. Clearly, the variety of the data may be great, the volume of data may be big, the collection velocity of data may be high, and the veracity of data may be problematic; yet, handled appropriately, the value can be significant.

Previous studies often focused on handling complexity in terms of computation-intensive operations. The focus has now switched to handling complexity in terms of data-intensive operations. In this respect, studies are carried out on every front. Notably, there are studies from the system perspective. These studies address the handling of big data at the processor level, at the physical machine level, at the cloud virtualization level, and so on. There are studies on data networking for big data communications and transmissions. There are also studies on databases to handle fast indexing, searches, and query processing. In the system perspective, the objective is to ensure efficient data processing performance, with trade-offs on load balancing, fairness, accuracy, outliers, reliability, heterogeneity, service level agreement guarantees, and so on.

Nevertheless, with the aforementioned real world applications as the demand, and the advances of the storage, system, networking and database support as the supply, their direct marriage may still result in unacceptable performance. As an example, smart sensing devices, cameras, and meters are now widely deployed in urban areas. Frequent checking needs to be made of certain properties of these sensor data. The data is often big enough that even process each piece of the data just once can consume a great deal of time. Studies from the system perspective usually do not provide an answer to the issue of which data should be processed (or given higher priority in processing) and which data may be omitted (or given a lower priority in processing). Novel algorithms, optimizations, and learning techniques are thus urgently needed in data analytics to wisely manage the data.

From a broader perspective, data and the knowledge discovery process involve a cycle of analyzing data, generating a hypothesis, designing and executing new experiments, testing the hypothesis, and refining the theory. Realizing the transformative potential of big data requires many challenges in the management of data and knowledge to be addressed, computational methods for data analysis to be devised, and many aspects of data-enabled discovery processes to be automated. Combinations of computational, mathematical, and statistical techniques, methodologies and theories are needed to enable these advances to be made. There have been many new advances in theories and methodologies on data analytics, such as sparse optimization, tensor optimization, deep neural networks (DNN), and so on. In applying these theories and methodologies to the applications, specific application requirements can be taken into consideration, thus wisely reducing, shaping, and organizing the data. Therefore, the final processing of data in the system can be significantly more efficient than if the application data had been processed using a brute force approach.

An overall picture of the big data processing is given in Fig. 1.2. At the top are real world applications, where specific applications are designed and the data are collected. Appropriated algorithms, theories, or methodologies are then applied to assist knowledge discovery or data management. Finally, the data are stored and processed in the execution systems, such as Hadoop, Spark, and others.

In this book, we specifically study one big data analytic theory, the *sublinear algorithm*, and its use in real-world big data applications. As the name suggested, the

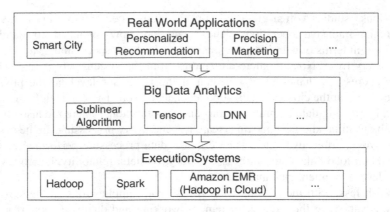

Fig. 1.2 A overall picture: from real world applications to big data analytics to execution systems

performance of the sublinear algorithms, in terms of time, storage space, and so on, is less than linear as against the amount of input data. More importantly, sublinear algorithms provide guarantees of accuracy of the output from the algorithms.

1.2 Sublinear Algorithms

Research on sublinear algorithms began some time ago. Sublinear algorithms were initially developed in the theoretical computer science community. The sublinear algorithm is one further classification of the approximation algorithm. Its study involves the long-debated issue of the trade-off between algorithm processing time and algorithm output quality.

In a conventional approximation algorithm, the algorithm can output an approximate result that deviates from the optimal result (within a bound), yet the algorithm processing time can become faster. One hidden implication of the design is that the approximate result is 100 % guaranteed within this bound. In a sublinear algorithm, such an implication is relaxed. More specifically, a sublinear algorithm outputs an approximate result that deviates from the optimal result (within a bound) for a (usually) majority of the time. As a concrete example, a sublinear algorithm usually says that the output of the algorithm differs from the optimal solution by at most 0.1 (the bound) at least 95 % of the time (the confidence).

This transition is important. From the theoretical research point of view, a new category is developed. From the practical point of view, sublinear algorithms provide two controlling parameters for the user in making trade-offs, while approximation algorithms have only one controlling parameter.

As can be imagined, sublinear algorithms are developed based on random and probabilistic techniques. Note, however, that the guarantee of a sublinear algorithm is on the individual outputs of this algorithm. In this, the sublinear algorithm differs from stochastic techniques, which analyze the mean and variance of a system in a steady state. For example, a typical queuing theory result is that the expected waiting time is 100 s.

In the theoretical computer sciences in the past few years, there have been many studies on sublinear algorithms. Sublinear algorithms have been developed for many classic computer science problems, such as finding the most frequently element, finding distinct elements, etc.; and for graph problems, such as finding the minimum spanning tree, etc.; and for geometry problems, such as finding the intersection of two polygons, etc. Sublinear algorithms can be broadly classified into sublinear time algorithms, sublinear space algorithms, and sublinear communication algorithms, where the amount of time, storage space, or communications needed is $o(N)$ with N as the input size.

Sublinear algorithms are a good match of big data analytics. Decisions can be drawn by only looking at a subset of the data. In particular, sublinear algorithms are suitable for situations, where the total amount of data is so massive that even linear processing time is not affordable. Sublinear algorithms are also suitable for situations, where some initial investigations need to be made before looking into the full data set. In many situations, the data are massive but it is not known whether the value of the data is big or not. As such, sublinear algorithms can serve to give an initial "peek" of the data before more a in-depth analysis is carried out. For example, in bioinformatics, we need to test whether certain DNA sequences are periodic. Sublinear algorithms, when appropriately designed to test periodicity in data sequences, can be applied to rule out useless data.

While there have been decent advances in the past few years in research on sublinear algorithms, to date, the study of sublinear algorithms has often been restricted to the theoretical computer sciences. There have been some applications. For example, in databases, where sublinear algorithms are used for the efficient query processing such as top-k queries; in bioinformatics, sublinear algorithms are used for testing whether a DNA sequence shows periodicity; and in networking, sublinear algorithms are used for testing whether two network traffic flows are close in distribution. Nevertheless, sublinear algorithms have yet to be applied, especially in the form of current big data applications. Tutorials on sublinear algorithms from the theoretical point of view, with a collection of different sublinear algorithms, aimed at better approximation bounds, are particularly abundant [2]. Yet there are far fewer applications of sublinear algorithms, aimed at application background scenarios, problem formulations, and evaluations of parameters. This book is not a collection of sublinear algorithms; rather, the focus is on the application of sublinear algorithms.

In this book, we start from the foundations of the sublinear algorithm. We discuss approximation and randomization, the later being the key to transforming a conventional algorithm to a sublinear one. We progressively present a few examples, showing the key ingredients of sublinear algorithms. We then discuss how to apply sublinear algorithms in three state-of-the-art big data domains, namely, data collection in wireless sensor networks, big data processing using MapReduce, and behavior analysis using metering data from smart grids. We show how the problem should be formalized, solved, and evaluated, so that the sublinear algorithms can be used to help solve real-world problems.

1.3   Book Organization

The purpose of this book is to give a snapshot of sublinear algorithms and their applications. Different from other writings on sublinear algorithms, we focus on learning the basic ideas of sublinear algorithms, rather than on presenting a comprehensive survey of the sublinear algorithms found in literature. We also target the issue of when and how to apply sublinear algorithms to applications. This includes learning in what situations the sublinear algorithms may fit into certain scenarios, how we may combine multiple sublinear algorithms to solve a problem, how to develop sublinear algorithms with additional statistical information, what structures are needed to support sublinear algorithms, and how we may extend existing sublinear algorithms to fit into applications. The remaining five chapters of the book are organized as follows.

In Chap. 2, we present the basic concepts of the sublinear algorithm. We first present the main thread of theoretical research on sublinear algorithms and discuss how sublinear algorithms are related to other theoretical developments in the computing sciences, in particular, approximation and randomization. We then present preliminary mathematical techniques on inequalities and bounds. We then give three examples. The first is on estimating the percentage of households among a group of people. This is an illustration of the direct application of inequalities and bounds to derive a sublinear algorithm. The second is on finding distinct elements. This is a classical sublinear algorithm. The example involves some key insights and techniques in the development of sublinear algorithms. The third is a two cat problem where we develop an algorithm that is sublinear, but which does not fall into standard sublinear algorithm format. The example provides some additional thoughts on the wide spectrum of sublinear algorithms.

In Chap. 3, we present an application of sublinear algorithms in wireless sensor data collection. Data collection is one of the most important tasks for a wireless sensor network. We first present the background in wireless sensor data collection. One problem of data collection arises when the total amount of data collected is big. We show that sublinear algorithms can be used to substantially reduce the number of sensors involved in the data collection process, especially when there is a need for frequent property checking. Here, we develop a layered architecture that can facilitate the use of sublinear algorithms. We then show how to apply and combine multiple sublinear algorithms to collectively achieve a certain task. Furthermore, we show that we can use side statistical information to further improve the performance.

In Chap. 4, we present an application of sublinear algorithms for big data processing in MapReduce. MapReduce, initially proposed by Google, is a state-of-the-art framework for big data processing. We first present the background of big data processing, MapReduce, and a data skew problem within the MapReduce framework. We show that the overall problem is a load balancing problem, and we formulate the problem. The problem calls for the use of an online algorithm. We first develop a straightforward online algorithm and prove that it is 2-competitive. We then show that by sampling a subset of the data, we can make wiser decisions. We develop an algorithm and analyze the amount of data that we need to "peek" before we can make theoretical guaranteed decisions. Intrinsically, this is a sublinear algorithm. In this application, the sublinear algorithm is not the solution for the entire problem space. We show that the sublinear algorithm assists in solving a data skew problem so that the overall solution is a more accurate one.

In Chap. 5, we present an application of sublinear algorithms for a behavior analysis using metering data from a smart grid. Smart meters are now widely deployed where it is possible to collect fine-grained data on the electricity usage of users. One objective is to conduct a classification of the users based on data of their electricity use. We choose to use the electricity usage distribution as the criterion for classification, as it captures more information on the behavior of a user. Such classification can be used for customized differentiated pricing, energy conservation, and so on. In this chapter, we first present a trace analysis on the smart metering data that we collected, which were recorded for 2.2 million households in the great Houston area. For each user, we recorded the electricity used every 15 min. Clearly, we face a big data problem. We develop a sublinear algorithm, where we apply an existing sublinear algorithm that was developed in the literature as a sub-function. Finally, we present differentiated services for a utility company. This shows a possible case of the use of user classifications to maximize the revenue of the utility company.

In Chap. 6, we present some experiences in the development of sublinear algorithms and a summary of the book. We discuss the fitted scenarios and limitations of sublinear algorithms as well as the opportunities and challenges to the use of sublinear algorithms. We conclude that there is an urgent need to apply the sublinear algorithms developed in the theoretical computer sciences to real-world problems.

References

1. Google Flu Prediction, available at http://www.google.org/flutrends/.
2. R. Rubinfeld, Sublinear Algorithm Surveys, available at http://people.csail.mit.edu/ronitt/sublinear.html.
3. Y. Zheng, F. Liu, and H. P. Hsieh, "U-Air: When Urban Air Quality Inference meets big Data", in *Proc. ACM SIGKDD'13*, 2013.
4. Y. Zhou, M. Tang, W. Pan, J. Li, W. Wang, J. Shao, L. Wu, J. Li, Q. Yang, and B. Yan, "Bird Flu Outbreak Prediction via Satellite Tracking", in *IEEE Intelligent Systems*, Apr. 2013.

Chapter 2
Basics for Sublinear Algorithms

2.1 Introduction

In this chapter, we study the theoretical foundations of sublinear algorithms. We discuss the foundations of approximation and randomization and show the history of the development of sublinear algorithms in the theoretical research line. Intrinsically, sublinear algorithms can be considered as one branch of approximation algorithms with confidence guarantees. A sublinear algorithm says that the accuracy of the algorithm output will not deviate from an error bound and there is high confidence that the error bound will be satisfied. More rigidly, a sublinear algorithm is commonly written as $(1 + \epsilon, \delta)$-approximation in a mathematical form. Here ϵ is commonly called an *accuracy parameter* and δ is commonly called a *confidence parameter*. This accuracy parameter is the same to the approximate factor in approximation algorithms. This confidence parameter is the key trade-off where the complexity of the algorithm can reduce to sublinear. We will rigidly define these parameters in this chapter.

Then we present some inequalities, such as Chernoff inequality and Hoeffding inequality, which are commonly used to derive the bounds for the sublinear algorithms. We further present the classification of sublinear algorithms, namely sublinear algorithms in time, sublinear algorithms in space, and sublinear algorithms in communication.

Three examples will be instanced in this chapter to illustrate how sublinear algorithms (in particular, the bounds), which are developed from the theoretical point of view. The first example is a straightforward application of Hoeffding inequality. The second one is a classic sublinear algorithm to find distinct elements. In the third example, we show a sublinear algorithm that does not belong to the standard form of (ϵ, δ) approximation. This can further broaden the view on sublinear algorithms.

© The Author(s) 2015
D. Wang, Z. Han, *Sublinear Algorithms for Big Data Applications*,
SpringerBriefs in Computer Science, DOI 10.1007/978-3-319-20448-2_2

2.2 Foundations

2.2.1 Approximation and Randomization

We start by considering algorithms. An algorithm is a step-by-step calculating procedure for solving a problem and outputting a result. In common sense, an algorithm tries to output an optimal result. When evaluating an algorithm, an important metric is its complexity. There are different complexity classes. Two most important classes are P and NP. The problems in P are those that can be solved in polynomial times and the problems in NP are those that must be solved in super-polynomial times. Using today's computing architecture, running polynomial time algorithms is considered tolerable within their finishing times.

To handle the problems in NP, a development from theoretical computer science is to introduce a trade-off where we sacrifice the optimality of the output result so as to reduce the algorithm complexity. More specifically, we do not need to achieve the exact optimal solution; yet it is acceptable if we know that the output is close to the optimal solution. This is called *approximation*. Approximation can be rigidly defined. We show one example on a $(1 + \epsilon)$-approximation.

Let Y be a problem space and $f(Y)$ be the procedure to output a result. We call an algorithm a $(1 + \epsilon)$-approximation if this algorithm returns $\hat{f}(Y)$ instead of the optimal solution $f^*(Y)$, and

$$|\hat{f}(Y) - f^*(Y)| \leq \epsilon f^*(Y)$$

Two comments have been made here. First, there can be other approximation criteria beyond $(1 + \epsilon)$-approximation. Second, approximation, though introduced mostly for NP problems, is not restricted to NP problems. One can design an approximation algorithm for the problems in P to further reduce the algorithm complexity as well.

A hidden assumption of approximation is that an approximation algorithm requests that its output is always, i.e., 100 %, within an ϵ factor of the optimal solution. A further development from theoretical computer sciences is to introduce another trade-off between optimality and algorithm complexity; that is, it is acceptable that the algorithm output is close to the optimal most of the times. For example, 95 % of time, the output result is close to the optimal result. Such probabilistic nature requires an introduction of *randomization*. We call an algorithm a $(1 + \epsilon, \delta)$-approximation if this algorithm returns $\hat{f}(Y)$ instead of the optimal solution $f^*(Y)$, and

$$Pr[|\hat{f}(Y) - f^*(Y)| \leq \epsilon f^*(Y)] \geq 1 - \delta$$

Here ϵ is usually called as an *accuracy parameter* (error bound) and δ is usually called as a *confidence parameter*.

Discussion: We have seen two steps in theoretical computer sciences in trading-off optimality and complexity. Such trade-off does not immediately lead to an algorithm that is sublinear to its input, i.e., $(1 + \epsilon, \delta)$-approximation is not necessarily sublinear. Nevertheless, these provide better categorization on algorithms. In particular, the second advancement in randomization makes a sublinear algorithm possible. As discussed in the introduction, processing the full data may not be tolerable in the big data era. As a matter of fact, practitioners have already designed many schemes using only partial data. These designs may be ad hoc in nature and may not have rigid proofs in their quality. Thus, from a quality-control's point of view, the $(1 + \epsilon, \delta)$-approximation brings to the practitioners a rigid theoretical evaluation benchmark when evaluating their designs.

2.2.2 Inequalities and Bounds

One may recall that the above formulas are similar to those inequalities in probability theory. The difference is that the above formulas and bounds are used on algorithms and in probability theory, the formulas and bounds are used on variables.

In reality, many developments of sublinear algorithms heavily apply probability inequalities. Therefore, we state a few mostly used inequalities here and we will use examples to show how they will be applied to sublinear algorithm development.

Markov inequality: For a nonnegative random variable X, and any $a > 0$, we have

$$Pr[X \geq a] \leq \frac{E[X]}{a}$$

Markov inequality is a loose bound. The good thing is that Markov inequality requires no assumptions on the random variable X.

Chernoff inequality: For independent random Bernoulli variables X_i, let $X = \sum X_i$. For any Δ, we have

$$Pr[X \leq (1 - \Delta)E[X]] \leq e^{-\frac{E[X]\Delta^2}{2}}$$

Chernoff bound is tighter. Note, however, that it requires the random variables to be independent.

Discussion: From probability theory, the intuition of Chernoff inequality is very simple. It says that the probability of the value of a random variable deviating from its expectation decreases very fast. From the sublinear algorithm point of view, the insight is that if we develop an algorithm and run this algorithm many times upon different subsets of randomly chosen partial data, the probability that the output of the algorithm deviating from the optimal solution decreases very fast. This is also called a *median trick*. We will see more on how to materialize this insight using examples throughout this book.

Chernoff inequality has many variations. Practitioners may often encounter a problem of computing $Pr[X \leq k]$ where k is a parameter of real world importance. Especially, one may want to link k with δ. For example, given that the expectation of X is known, how can the k be determined so that the probability $Pr[X \leq k]$ is at least $1 - \delta$. Such linkage between k and δ can be derived from Chernoff inequality as follows:

$$Pr[X \leq k] = Pr[X \leq \frac{k}{E[X]}E[X]]$$

Let $1 - \Delta = \frac{k}{E[X]}$ and with Chernoff inequality we have:

$$Pr[X \leq k] \leq e^{-\frac{E[X](1-\frac{k}{E[X]})^2}{2}}$$

Then, to link δ and k, we have

$$Pr[X \leq k] \leq e^{-\frac{E[X](1-\frac{k}{E[X]})^2}{2}} \leq 1 - \delta$$

Note that the last inequality provides a connection between k and δ.

Chebyshev inequality: For any X with $E[X] = \mu$ and $Var[X] = \sigma^2$, and for any $a > 0$,

$$Pr[|X - \mu| \geq a\sigma] \leq \frac{1}{a^2}$$

Hoeffding inequality: Assume we have k random identical and independent variables X_i, for any ϵ, we have

$$Pr[|X - E[X]| \geq \epsilon] \leq e^{-2\epsilon^2 k}$$

Hoeffding inequality is commonly used to bound the deviation from the mean.

2.2.3 Classification of Sublinear Algorithms

The most common classification of sublinear algorithms is to see whether a sublinear algorithm uses $o(N)$ in space or $o(N)$ in time or $o(N)$ in communication, where N is the input size. Respectively, they are called sublinear algorithms in time, sublinear algorithms in space or sublinear algorithms in communication.

Sublinear algorithms in time mean that one needs to make decisions yet it is impossible for him to look at all data; note that it takes a linear amount of time to look at all data. The result of the algorithm is using $o(N)$ time, where N is the input size. Sublinear algorithms in space mean that one can look at all data because the data is coming in a streaming fashion. In other words, the data comes in an online fashion and it is possible to read each piece of data as time progresses. Yet the challenge is that it is impossible to store all these data in storage because the data is too large. The result of the algorithm is using $o(N)$ space, where N is the storage space. Such category is also commonly called as data stream algorithms. Sublinear algorithms in communication mean that the data is too large to be stored in a single machine and one needs to make decision through collaboration between machines. It is only possible to use $o(N)$ communications, where N is the total number of communications.

There are algorithms that do not fall into the $((1+\epsilon), \delta)$-approximation category. A typical example is when there needs of a balance between the resources such as storage, communications, and time. Therefore, algorithms can be developed where the contribution of each type of resources is sublinear; and they collectively achieve the task. One example of such kind can be found from a sensor data collection application in [2]. In this example, a data collection task is achieved with a sublinear sacrifice of storage and a sublinear sacrifice of communication.

In this chapter, we will present a few examples. The first one is a simple example on estimating percentage. We show how the bound of a sublinear algorithm can be derived using inequalities. This is a sublinear algorithm in time. Then we discuss a classic sublinear algorithm to find distinct elements. The idea is to see how we can go beyond simple sampling and quantify an idea and develop quantitative bounds. In this example, we also show the median trick, a classic trick in managing δ. This is a sublinear algorithm in space. Finally, we discuss a two-cat problem, where its intuition is applied in [2]. This divides two resources and collectively achieves a task.

2.3   Examples

2.3.1   Estimating the User Percentage: The Very First Example

We start from a simple example. Assume that there is a group of people, who can be classified into different categories. One category is the housewife. The question is that we want to know the percentage of the housewife in this group, but the group is too big to examine every person. A simple way is to sample a subset of people and see how many of these people in it belong to the housewife group. This is where the question arise: how many samples are enough?

Assume that the percentage of housewife in this group of people is α. We do not know α in advance. Let ϵ be the error allowed to deviate from α and δ be a confidence interval. For example, if $\alpha = 70\%$, $\epsilon = 0.05$ and $\delta = 0.05$, it means that we can output a result where we have a 95 % confidence/probability that this result falls in the range of 65–75 %. The following theorem states the number of samples k we need and its relationship with ϵ, δ.

Theorem 2.1. *Given ϵ, δ, to guarantee that we have a probability of $1 - \delta$ success that the percentage (e.g., of housewife) will not deviate from α for more than ϵ, the number of users we need to sample must be at least $-\frac{\log \delta}{2\epsilon^2}$.*

We first conduct some analyses. Let N be the total number of users and let m be the number of users we sample. Let Y_i be an indicator random variable where

$$Y_i = \begin{cases} 1, & \text{housewife} \\ 0, & \text{otherwise} \end{cases}$$

We assume that Y_i are independent, i.e., Alice belongs to the housewife group is independent of whether Mary belongs to housewife or not.

Let $Y = \sum_{i=1}^{N} Y_i$. By definition, we have $\alpha = \frac{1}{N} E[Y]$. Since Y_i are all independent, $E[Y_i] = \alpha$. Let $X = \sum_{i=1}^{m} Y_i$. Let $\overline{X} = \frac{1}{m} X$. The next lemma says that the expectation $\overline{X}$ of the sampled set is the same as the expectation of the whole set.

Lemma 2.1. $E[\overline{X}] = \alpha$.

Proof. $E[\overline{X}] = \frac{1}{m} E[\sum_{1}^{m} Y_i] = \frac{1}{m} \times m\alpha = \alpha$. □

We next proof Theorem 2.1.

Proof.

$$Pr[(\overline{X} - \alpha) > \epsilon] = Pr[(\overline{X} - E[\overline{X}]) > \epsilon] \leq e^{-2\epsilon^2 m}$$

The last inequality is derived by Hoeffding Inequality. To make sure that $e^{-2\epsilon^2 m} < \delta$, we need to have $m > -\frac{\log \delta}{2\epsilon^2}$. □

Discussion: Sampling is not a new idea. Many practitioners naturally use sampling techniques to solve their problems. Usually, practitioners discuss the expected values, which ends up with a statistical estimation. In this example, the key idea is to transform a statistical estimation of the expected value into a bound.

2.3.2 Finding Distinct Elements

We now study a classic problem by using sublinear algorithms. We want to count the total number of distinct elements in a data stream. For example, suppose that we have a data stream $S = \{1, 2, 3, 1, 2, 3, 1, 2\}$. Clearly, the total number of distinct elements in S is 3.

We look for an algorithm that is sublinear in space. This means that at any single point of time, only a subset of elements can be stored in the memory. The algorithm will go over one pass of the data stream. Our algorithm will only store $O(\log N)$ data, where N is the total number of elements.

2.3.2.1 The Initial Algorithm

Let the number of distinct elements in S be F. Let $w = \log N$. Assume we have a hash function $h(\cdot)$, which can uniformly hash an element k into $[0, 2^w - 1]$. Let $r(\cdot)$ be a function that calculates the trailing 0's (counting from the right) in the binary representation of $h(\cdot)$. Let $R = \max r(\cdot)$.

We explain these notations through examples. Consider the above stream S. A hash function can be $h(k) = 3k + 1 \mod 8$. Then S is transformed into 4, 7, 1, 4, 7, 1, 4, 7. The $r(h(k))$ is then 2, 0, 0, 2, 0, 0, 2, 0. Hence, $R = 2$.

The algorithm is shown in Algorithm 1. We only need to store R, and clearly, R can be stored in $w = O(\log N)$ bits.

Algorithm 1 Finding Distinct Elements

Input: S. **Output:** $\hat{F}$.
  1: **for** Read each element k in S; **do**
  2:     Calculate $r(h(k))$;
  3:     Update R;
  4:     Discard k;
  5: Output $\hat{F} = 2^R$;

Still using our example of S where $R = 2$, the output result is $\hat{F} = 2^2 = 4$. This is an approximate to the true result $F = 3$.

This algorithm is not a direct application of sampling. The basic idea is as follows. The first step is to map the elements uniformly in the range of $[0, 2^w - 1]$. This avoids the problem that some elements are clustered in a small range. The second step is to convert each of the mapping results into the number of zeros starting counting from the right (counting from the left has a similar effect). Intuitively, if the number of distinct elements is big, there is a greater probability that such hashing hits a number with more zeros starting counting from the right.

Next, we analyze this algorithm. The next theorem shows that the approximate $\hat{F}$ is neither too big (overestimate), nor too small (underestimate) as compared to F.

Theorem 2.2. *For any integer $c > 2$, the probability that $\frac{1}{c} \leq \frac{\hat{F}}{F} \leq c$ is at least $1 - \frac{2}{c}$.*

We need a set of lemma before finally proving this theorem. First, next lemma states that the probability that we will hit a $r(h(k))$ with a large number of trailing 0s is exponentially decreasing.

Lemma 2.2. *For any integer $j \in [0, w]$, $Pr[r(h(k)) \geq j] = \frac{1}{2^j}$.*

Proof. Intrinsically, we are looking for $\underbrace{1\ldots1}_{w-j}\underbrace{0\ldots0}_{j}$. Since the hashing makes the elements of $h(k)$ uniformly distributed in $[0, 2^w]$, we have $Pr[r(h(k)) \geq j] = \frac{1}{2^j}$. $\quad\square$

Now we consider that the approximate $\hat{F}$ is an overestimation or an underestimation respectively.

We start from bounding that $\hat{F}$ is an overestimation. More specifically, given a constant c, we do not want $\hat{F}$ to be greater than cF.

Let Z_j be the number of distinct items in the stream S for which $r(f(k)) \geq j$. We are interested in the smallest j such that $2^j > cF$. If we do not want an overestimation, this Z_j should not be big, because if $Z_j = 1$, our output will be at least 2^j. Next lemma states that this is indeed true. In other words, the probability that $Z_j \geq 1$ can be bounded.

Lemma 2.3. $Pr[Z_j \geq 1] \leq \frac{1}{c}$.

Proof. Clearly, Z_j is an indicator variable such that

$$Z_j = \begin{cases} 1, & \text{if } r(f(k)) \geq j \\ 0, & \text{otherwise} \end{cases}$$

Thus,

$$E[Z_j] = F \times Pr[r(h(k)) \geq j] = \frac{F}{2^j}$$

by Markov inequality, we have

$$Pr[Z_j \geq 1] \leq E[Z_j]/1$$

Therefore,

$$Pr[Z_j] \geq 1 \leq E[Z_j]/1 = \frac{F}{2^j} \leq \frac{1}{c}$$

and this completes the proof.   □

We now look that the approximate $\hat{F}$ is an underestimation. More specifically, given a constant c, we do not want $\hat{F}$ that is less than $\frac{F}{c}$.

Again, let Z_l be the number of distinct items in the stream S for which $r(f(k)) \geq l$. We are interested in the smallest l such that $2^l < \frac{F}{c}$. If we do not want an underestimation, this Z_l should be at least 1, because if $Z_l = 0$, our output will be less than 2^l. Next lemma states that this is indeed true. In other words, the probability that $Z_l = 0$ can be bounded.

Lemma 2.4. $Pr[Z_l = 0] \leq \frac{1}{c}$.

Proof. Clearly, and again, Z_l is an indicator variable such that

$$Z_l = \begin{cases} 1, & \text{if } r(f(k)) \geq l \\ 0, & \text{otherwise} \end{cases}$$

Thus,

$$E[Z_l] = F \times Pr[r(h(k)) \geq l] = \frac{F}{2^l}$$

and

$$Var[Z_l] = F\frac{1}{2^l}(1 - \frac{1}{2^l})$$

$Pr[Z_l = 0] = Pr[Z_l - E[Z_l]] \geq E[Z_l]$ (assigning $Z_l = 0$ in the right hand side)

$$\leq \frac{Var[Z_l]}{E[Z_l]^2} \text{ (by Chebyshev inequality)}$$

$$< \frac{E[Z_l]}{E[Z_l]^2} \text{ (see } E[Z_l] \text{ and } Var[Z_l] \text{ developed above)}$$

$$= \frac{1}{E[Z_l]} = \frac{2^l}{F} < \frac{1}{c}.$$

and this completes the proof. □

By Lemma 2.3 and Lemma 2.4, we will not overestimate and underestimate a combined probability of more than $\frac{2}{c}$. We have thus proved Theorem 2.2.

2.3.2.2 Median Trick in Boosting Confidence

Algorithm 1 can output an approximate $\hat{F}$ of the true F with a constant probability. In reality, we may want the probability to be arbitrarily close to 1. A common trick to do this, i.e., boost the success probability, is called the median trick.

The algorithm is shown in Algorithm 2.

Algorithm 2 Final Algorithm

1: Run t copies of Algorithm 1 using mutually independent random has functions;
2: Output the median of the t answers;

The next theorem states that t can be as small as $\log \frac{1}{\delta}$. Thus, the total storage required is $O(\log \frac{1}{\delta} \log N)$.

Theorem 2.3. *There is a $t = O(\log \frac{1}{\delta})$ ensuring that $\frac{F}{c} \leq F \leq cF$ happens with probability at least $1 - \delta$.*

Proof. Define $x_i = 0$ if $\frac{\hat{F}}{F}$ is in $[\frac{1}{c}, c]$ or 1 otherwise. Let $X = \sum_{i=1}^{t} x_i$.

Note that we can associate x_i with each copy of the algorithm running in parallel and X indicates the total number of failure. Because we will output the median, we fail only if more than half of the parallel-running algorithms fail. In other words, $X > \frac{t}{2}$. Our objective is to find a t that this happens with very small probability δ.

From another angle, we want $X < \frac{t}{2}$ to be $1 - \delta$.

$$Pr[X \leq \frac{t}{2}] \leq e^{-\frac{E[X](1-\frac{t/2}{E[X]})^2}{2}} \leq (1 - \delta)$$

We know that $E[x_i] = \frac{2}{c}$ from Theorem 2.2. Thus $E[X] = t\frac{2}{c}$.
To solve this inequality, we have

$$t \geq c(1 - \frac{c}{4})^2 \log \delta$$

and this completes the proof. $\square$

Discussion: The bound in this example is not tight. We use c instead of ϵ as c is an integer constant. There are other bounds for finding distinct elements. Nevertheless, our goal is to show some core development methods for sublinear algorithms. Most notably, we see how to develop bounds given some key insights. Again this is related to the fact that the probability that deviates from the expectation can be bounded and the variance can be bounded. The median trick is a common trick to further boost the probability. In addition, one may find that sublinear algorithms are very simple in implementation, yet the analysis is usually complex.

2.3.3 Two-Cat Problem

We now study one problem that does not fall in the form of $(1 + \epsilon, \delta)$ approximation. Yet, the problem can be solved in a sublinear amount of resources. The problem is as follows.

The Two-Cat Problem: Consider a tall skyscraper building and you do not know the total number of floors of this building. Assume you have cats. The cats have the following properties: when you throw a cat out of window from floor $1, 2, \cdots, n$, the cat will survive; and when you throw a cat out of window from floor $n + 1, n + 2, \cdots, \infty$, the cat will die. The question is to develop an efficient method to determine n given that you have *two* cats.

We first look at when there is only one cat. It is easy to see that we have to use this cat to test floor one by one from $1, 2, \cdots, n$. This will take a total of n tests, which is linear. This n is also a lower bound.

Now we see the case that we have two cats. Before we give out the final solution, we first analyze the algorithm of an exponential increase algorithm. The algorithm is shown in Algorithm 3.

Algorithm 3 Exponential Increase Algorithm

1: $i = -1$
2: **for** the first cat is still alive **do**
3:     $i + +$;
4:     test floor 2^i;
   $k = 0$;
5: **for** the second cat is still alive **do**
6:     $k + +$;
7:     test floor $2^{i-1} + k$;
8: Output $2^{i-1} + k$;

The first cat in this algorithm will be used to test floors of $1, 2, 4, 8, 16, \cdots ,$. Assume that the first cat die on floor l; then the second cat will be used to test floors $2^{(\log l - 1)} + 1, 2^{(\log l - 1)} + 2, \cdots , l - 1, l$. For example, assume that $n = 23$. Using the above exponential algorithm, the first cat survives when it is used to test floor 16 and dies when it is used to test floor 32. Then we use the second cat to test floor 16–32. We finally conclude that $n = 23$.

It is easy to see that this exponential algorithm also takes linear time. The first cat takes $O(\log n)$ time, and the second cat, where the primary complexity comes from, takes $O(\frac{n}{2})$. This leads to a linear complexity to the overall algorithm.

Now we present a sublinear algorithm in Algorithm 4.

Algorithm 4 The Two-Cat Algorithm

1: $i = -1, l = 1$
2: **for** the first cat is still alive **do**
3:     $i + +$;
4:     test floor $l = l + i$;
   $l = l - i$;
5: **for** the second cat is still alive **do**
6:     $l + +$;
7:     test floor l;
8: Output l;

The first cat in this algorithm will be used to test floors of $1, 3, 6, 10, 15, 21, 28, \cdots ,$. Assume that the first cat dies on l at the ith round; then the second cat will be used to test floors $l - i + 1, l - i + 2, \cdots , l - 1, l$. For example, assume that $n = 23$. Using the above algorithm, the first cat survives when it is used to test floor 21 and dies when it is used to test floor 28. Then we use the second cat to test floor 21–28 floor by floor, and conclude that $n = 23$.

Now we analyze this algorithm.

Lemma 2.5. *The Two-Cat Algorithm is sublinear and it takes $O(\sqrt{n})$ steps.*

Proof. Sketch: Assume the first cat takes x step and the second one takes y steps.

For the first cat, the final floor l it reaches is equal to $1 + 2 + 3 + 4 + \cdots$. More specifically, we have $l = \frac{(1+x)x}{2}$. Clearly, $l = O(n)$. Thus, $x = O(\sqrt{n})$.

For the second cat, we look at the total number of floors l' before the first cat dies. This is $l' = \frac{x(x-1)}{2}$. The maximum number of floors this second cat should test is equal to $l - l'$, i.e., $y = O(l - l')$. Therefore, $y = O(\frac{(1+x)x}{2} - \frac{x(x-1)}{2}) = O(x)$. Hence, $y = O(\sqrt{n})$.

Combining x and y, we have the complexity of the algorithm $O(\sqrt{n})$. □

For this two-cat problem, $O(\sqrt{n})$ is also a lower bound, i.e., this algorithm is the fastest algorithm that we can gain. We omit the proof. One may be interested in investigating the case, if we have three cats. The result will be $O(\sqrt[3]{n})$.

Discussion: This algorithm has important indication. Intrinsic we may consider that the two cats are two pieces of resources. This problem shows that, to collectively achieve a certain task, we can divide the resources where each piece of the resource undertakes a sublinear overhead.

One work of Partial Network Coding (PNC) applies this idea [2]. In PNC, there are two pieces of resources, namely, communication and storage. To collectively achieve a task, either the total amount of communication needs to be linear or the total amount of storage needs to be linear. In [2], it is shown that we can divide the overhead into a $O(\sqrt{N})$ factor for the communication and a $O(\sqrt{N})$ factor to storage, so that each resource has a sublinear overhead.

2.4 Summary and Discussions

In this chapter, we present the foundations, some examples and common algorithm development techniques for sublinear algorithms. We first present how sublinear algorithms are developed from the theoretical point of view, in particular, its connections with approximation algorithms. Then we present a set of commonly used inequalities that are important for approximation bound development. We start from a very simple example that directly applies Hoeffding inequality. Then we study a classic example of sublinear algorithm to find distinct elements. Some commonly used tricks for boosting confidence are presented. Finally, we present a two-cat problem.

These examples reveal how sublinear algorithms are developed from theoretical point of view, i.e., most importantly, how the bounds are derived from. In the following chapters of applications of sublinear algorithms, we will see that sometimes we need to derive certain bounds given specific application scenarios. This means one may need to master theoretical algorithm development to certain extent. In other times, we will see that we can apply existing sublinear algorithms. This means that one may need to refer to existing theoretical development of sublinear algorithms. From the targets of this book on application of sublinear algorithms, it is clear that it is always better to know more existing sublinear algorithms in literature, consider [1] as a comprehensive survey.

References

1. R. Rubinfeld, Sublinear Algorithm Surveys, available at http://people.csail.mit.edu/ronitt/sublinear.html.
2. D. Wang, Q. Zhang, and J. Liu, "Partial Network Coding: Theory and Application in Continuous Sensor Data Collection", in *Proc. IEEE IWQoS'06*, New Haven, CT, Jun. 2006.

Chapter 3
Application on Wireless Sensor Networks

3.1 Introduction

Wireless sensor networks provide a model in which sensors are deployed in large numbers where traditional wired or wireless networks are not available/appropriate. Their intended uses include terrain monitoring, surveillance, and discovery [11] with applications to geological tasks such as tsunami and earthquake detection, military surveillance, search and rescue operations, building safety surveillance (e.g. for fire detection), and biological systems.

The major difference between sensor networks and traditional networks is that unlike a host computer or a router, a sensor is typically a tightly-constrained device. Sensors not only lack long life spans due to their limited battery power, but also possess little computational power and memory storage [1]. As a result of the limited capabilities of individual sensors, one sensor usually can only collect a small amount of data from its environment and carry out a small number of computations. Therefore, a single sensor is generally expected to work in cooperation with other sensors in the network. As a result of this unique structure, a sensor network is typically data-centric and query-based [8]. When a query is made, the network is expected to distribute the query, gather values from individual sensors, and compute a final value. This final value typically represents key properties of the area where the network is deployed; examples of such values are MAXIMUM, MINIMUM, QUANTILE, AVERAGE, and SUM [18, 19] over the individual parameters of the sensors, such as temperature and air or water composition. As an example, consider a sensor network monitoring the average vibration level around a volcano. Each sensor lying in the crater area submits its own value representing the level of activity in a small area around it. Then the data values are relayed through the network; in this process, they are aggregated so that fewer messages need to be sent. Ultimately, the base station obtains the aggregated information about the area being monitored.

In addition to their distributed nature, most sensor networks are highly redundant to compensate for the low reliability of the sensors and environmental conditions.

© The Author(s) 2015
D. Wang, Z. Han, *Sublinear Algorithms for Big Data Applications*,
SpringerBriefs in Computer Science, DOI 10.1007/978-3-319-20448-2_3

Since the data from a sensor network is the aggregation of data from individual sensors, the number of sensors in a network has direct influence on the delay incurred in answering a query. In addition, significant delay is introduced by in-network aggregation [14, 16, 18], since intermediate parent nodes have to wait for the data values collected from their children before they can associate them with their own data.

While most of the techniques for fast data gathering focus on delay-energy efficiencies, they lack provable guarantees for the accuracy of the result. In this chapter, we focus on a new approach to address the delay and accuracy challenges. We propose a simple distributed architecture which consists of layers, where each layer contains a subset of the sensor nodes. Each sensor randomly promotes itself into different layers, where large layers contain a superset of the sensors on smaller layers. The key difference between our layered architecture and hierarchical architectures is that each sensor in our network only represents itself and submits its own data to each query, without the need to act as a "head" of a cluster of sensors. In this model, a query will be made to a particular layer, resulting in an aggregation tree with fewer hops, and thus smaller delay. Unfortunately, the reduction in delay comes with a price tag; since only a subset of the sensors submit their data, the accuracy of the answer to the query is compromised.

In this chapter, we study the tradeoff between the delay and the accuracy with proving bounds. We implement this study in the context of five key properties of the network, MAX, MIN, QUANTILE, AVERAGE and SUM. Given a user defined accuracy level, we analyze what the layer of the network should be queried for these properties. We show that different queries do show distinct characteristics which affect the delay/accuracy tradeoff. Meanwhile, we present that for certain types of queries such as AVERAGE and SUM, additional statistical information obtained from the history of the environment can help further reduce the number of sensors involved in answering a query. We then investigate the new tradeoffs given the additional information.

The algorithm that we propose for our architecture is fully distributed; there is no need for the sensors to keep information about other sensors. Using the fact that each sensor is independent of others, we show how to balance the power consumption at each node by reconstructing the layered structure periodically, which results in an increase in the life expectancy of the whole network.

3.1.1 Background and Related Work

Wireless sensor networks have gained tremendous attention from the very beginning of its proposal. There is a wide range of applications; initially it starts from the wild and battlefield and recently moves to urban applications. The key advantages of a wireless sensor network is the wireless communication making it cheap and readily deployable; its self-organization nature; and its deep penetration to the physical environments. Some surveys on the challenges, techniques and protocols of wireless sensor networks can be found in [1, 2, 8].

One key objective of wireless sensor network is data collection. Different from data transmission of traditional networking, which is address-based and end-to-end, wireless sensor data collection is data centric, commonly integrated with in-network aggregation. More specifically, each individual sensor contributes its own data and the sensors of the whole network collectively achieve a certain task. There are many research issues related to sensor data collection, in particular, many focus on trade-off between key parameters, such as query accuracy, delay and energy usage (or load balancing).

SPIN [10] is the first data centric protocol which uses flooding; Directed Diffusion [13] is proposed to select more efficient paths. Several related protocols with similar concepts can be found in [5, 7, 20]. As an alternative to flat routing, hierarchical architectures have been proposed for sensor networks; in LEACH [11], heads are selected for clusters of sensors; they periodically obtain data from their clusters. When a query is received, a head reports its most recent data value. In [24], energy is studied in a more refined way in which a secondary parameter such as node proximity or node degree is included. Clustering techniques are studied in a different fashion in several papers, where [15] focuses on non-homogeneously dispersed nodes and [3] considers spanning tree structures. In-network data aggregation is a widely used technique in sensor networks [18, 19, 23]. Ordered properties, for example, QUANTILE are studied in [9]. A recent result in [6] considers power-aware routing and aggregation query processing together, building energy-efficient routing trees explicitly for aggregation queries.

Delay issues in sensor networks are mentioned in [16, 18] where the aggregation introduces high delay since each intermediate node and the source have to wait for the data values from the leaves of the tree, as confirmed by Yu et al. [25]. In [14], where a modified direct diffusion is proposed, a timer is set up for intermediate nodes to flush data back to the source if the data from their children have not been received within a time threshold. In case of energy-delay tradeoffs, [25] formulates delay-constraint trees. A new protocol is proposed in [4] for delay critical applications, in which energy consumption is of secondary importance. In these algorithms, all of the sensors in the network are queried, resulting in $\Theta(N)$ processing time, where N denotes the number of sensors in the network, which incurs long delay. Embedding hierarchical architectures into the network where a small set of "head" sensors collect data periodically from their children/clusters and submit the results that queried [11, 17, 24] provides a very useful abstraction, where the length of the period is crucial for the tradeoff between the freshness of the data and the overhead.

3.1.2 Chapter Outline

We present the system architecture in Sect. 3.2. Section 3.3 contains the theoretical analysis of the tradeoff between the accuracy of query answers and the latency of the system. In Sect. 3.4, we address the energy consumption of our system. Section 3.5 evaluates the performance of our system using simulations. We further present some variations of the architecture in Sect. 3.6. In Sect. 3.7, we summarize this application and how the sublinear algorithms are used in this application.

3.2 System Architecture

3.2.1 Preliminaries

We assume our network has N sensors denoted by $s_1, s_2, \ldots, s_N$ and deployed uniformly in a square area with side length D. We assume that a base station acts as an interface between the sensor network and the users, receiving queries which follow a Poisson distribution with the mean interval length λ.

We embed a layered structure in our network, with L layers, numbered 0, 1, 2, ..., $L - 1$. We use $r(l)$ to denote the transmission range used on layer l: during a transmission taking place on layer l, all sensors on layer l communicate by using $r(l)$ and can reach one another, in one or multiple hops. Let $e(l)$ be the energy needed to transmit for layer l. The energy spends per sensor for a transmission is $e(l) = r(l)^\alpha$ where $2 \leq \alpha \leq 4$ [22]. Initially, each sensor is at energy level B, which decreases with each transmission. R denotes the maximum transmission range of the sensors.

3.2.2 Network Construction

We would like to impose a layered structure on our sensor network where each sensor will belong to one or more layers. The properties of this structure is as follows.

(a) The *base layer* contains all sensors $s_1, \ldots, s_N$.
(b) The layers are numbered 0 through $L - 1$, with the base layer labelled 0.
(c) The sensors on layer l form a subset of those on layer $l - 1$, for $1 \leq l \leq L - 1$.
(d) The expected number of sensors on each layer drops exponentially with the layer number.

We now expound on how this structure is constructed. In our scheme, each sensor decides, without requiring any communication with the outside world, to which layer(s) it will belong. We assume that all the sensors have access to a value $0 < p < 1$ (this value may be hardwired into the sensors). Let us consider the decision

Fig. 3.1 A Layered Sensor
Network; a link is presented
whenever the sensor nodes in
a certain layer are within
transmission range

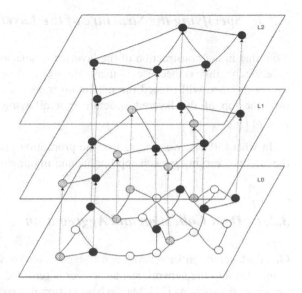

process that a generic sensor s_i undergoes. All sensors, including s_i, exist in the base
layer 0. Inductively, if s_i exists on some layer l, it will, with probability p, *promote*
itself to layer $l + 1$, which means that s_i will exist on layer $l + 1$ *in addition to* all
the lower layers $l, l - 1, , \ldots, 0$. If on some layer l', s_i makes the decision not to
promote itself to layer $l' + 1$, s_i stops the randomized procedure and does not exist
on any higher layers. If s_i promotes itself to the highest layer $L - 1$, it stops the
promotion procedure since no sensor is allowed to exist beyond layer $L - 1$. Thus,
any sensor will exist on layers $0, 1, \ldots, k$ for some $0 \le k \le L - 1$. Figure 3.1 shows
the architecture of a sensor network with three layers.

Since our construction does not assume the existence of any mechanism of
synchronization, it is possible that some sensors may be late in completing its
procedure for promoting itself up the layers. Since the construction scheme works
in a distributed fashion, this is not a problem—the late sensor can simply promote
itself using probability p and join its related layers in its own time.

Whenever the base station has a query, the query is sent to a specific Layer.
Those and only those sensors existing on this layer are expected to take place in
the communication. This can be achieved by reserving a small field (of $\log \log N$
bits) in the transmission packet for the layer number. Once l is specified by the
base station (the method for which will be explained later), all of the sensors on
layer l communicate using transmission range $r(l)$. The transmission range can be
determined by the expected distance of two neighboring sensors on layer l, i.e.
$r(l) = \frac{D}{\sqrt{N/2^l}}$, and can be enlarged a little further to ensure higher chances of
connectivity.

3.2.3 Specifying the Structure of the Layers

Note that in the construction of the layers, the sensors do not promote themselves
indefinitely; this is because if there are too few sensors on a layer, the inter-
sensor distance will exceed the maximum transmission range R. Rather, we "cut
off" the top of the layered structure, not allowing more than L layers where
$L = \Theta \left(\log \frac{N}{(\frac{D}{R}+1)^2} \right)$.

In what follows, we assume that the promotion probability $p = \frac{1}{2}$. We analyze
the effect of varying p when appropriate and in our simulations.

3.2.4 Data Collection and Aggregation

Given a layered sensor network constructed as above, we now focus on how a query
is injected into the network and an answer is returned. We simplify the situation by
assuming the same as [21] that the base station is a special node where a query will
be initiated. Thus the base station acts as an interface between the sensor network
and the user.

When the base station has a query to make, it first determines which layer is
to be used for this query. Let this layer be l. The base station then broadcasts the
query using communication range $r(l)$ for this layer. In this message, the base station
specifies the layer number l and the query type (in this chapter, we study MAX,
MIN, QUANTILE, AVERAGE and SUM). Any sensor on layer l that hears this
message will relay information using communication range $r(l)$; those sensors not
on layer l will simply ignore this message.

After the query is received by all the sensors on layer l, a routing tree rooted at
the base station is formed. Each leaf node then collects its data and sends it to its
parent, which then aggregates its own data with the data from its children, relaying
it up to its parent. Once the root has the aggregated information, it can obtain the
answer to the query.

Note that our schemes are independent of the routing and aggregation algorithms
used in the network. Our goal is to specify the layer number l which will reduce
the number of sensors, as well as the number of messages, used in responding to a
query. Once l is determined, the distribution of the query and the collection of the
data can be performed in a number of ways, such as that proposed in [6]. In fact,
once the layer to be used for a particular query has been identified, the particular
routing/aggregation algorithm to be used is transparent to our algorithm.

3.3 Evaluation of the Accuracy and the Number of Sensors Queried

In this section, we explore how the accuracy of the answers to queries and the latency relate to the layer which is being queried. In general, we would like to be able to obtain the answers to the queries with as little delay as possible. This delay is a function of the number of sensors whose data are being utilized for a particular query. Thus, the delay is reflected by the layer to which the query is sent. We would also like to get as accurate answers to our queries as possible. When a query utilizes data from all the sensors, the answer is accurate; however, when readings from only a subset of the sensors are used, errors are introduced. We now analyze how these concerns of delay and accuracy relate to the number of sensors queried, and thus to the layer used.

To explore the relation between the accuracy of the answer to a query and the layer l to which the query has been sent, we recall that the current configuration of the layers have been reached by each sensor locally, which decides how many layers it will exist. Due to the randomized nature of this process, the number of sensors on each layer is a random variable. In the next lemma, we investigate which layer must be queried if one would like to have input from at least k sensors.

Lemma 3.1. *Let* $l < \log N - \log\left(k + \ln\frac{1}{\delta} + \sqrt{\ln\frac{1}{\delta}(2k + \ln\frac{1}{\delta})}\right)$, *where* $k \leq$ *the expected number of sensors on layer* l. *Then, the probability that there are fewer than* k *sensors on layer* l *is less than* δ.

Proof. Define random variable Y_i for $i = 1, \ldots, N$ as follows. $Y_i = 1$ if s_i is promoted to layer l; and $Y_i = 0$ otherwise. Clearly, $Y_1, \ldots, Y_N$ are independent. $Pr[Y_i = 1] = 1/2^l$, and $Pr[Y_i = 0] = 1 - 1/2^l$. On layer l there are $Y = \sum_{i=1}^{N} Y_i$ sensors. Therefore, $Pr[Y < k] = Pr[Y < \frac{k}{E[Y]}E[Y]] < e^{-(1-\frac{k}{E[Y]})^2 E[Y]/2}$ by Chernoff's inequality. Since $E[Y] = N/2^l$, to have $e^{-(1-\frac{k}{E[Y]})^2 E[Y]/2} < \delta$, we must have $l < \log N - \log\left(k + \ln\frac{1}{\delta} + \sqrt{\ln\frac{1}{\delta}(2k + \ln\frac{1}{\delta})}\right)$ □

In what follows, we analyze the accuracy and the layer in the context of certain types of queries.

3.3.1 MAX and MIN Queries

In general, exact answers to maximum or minimum queries cannot be obtained unless all sensors in the network contribute to the answer, since any missed sensors might contain an arbitrarily high or low data value. The following theorem is immediate.

Theorem 3.1. *The queries for MAX and MIN must be sent to the base layer to avoid arbitrarily high error.*

3.3.2 QUANTILE Queries

As we cannot obtain an exact quantile by querying a proper subset of the sensors in the network we first introduce the notion of *an approximate number of quantile*.

Definition 3.1. The ϕ-*quantile* ($\phi \in (0, 1]$) of an ordered sequence S is the element whose rank in S is $\phi|S|$.

Definition 3.2. An element of an ordered sequence S is the ϵ-*approximation* ϕ-*quantile* of S if its rank in S is between $(\phi - \epsilon)|S|$ and $(\phi + \epsilon)|S|$.

The following lemma shows that a large enough subset of S has similar quantiles to S.

Lemma 3.2. *Let $Q \subseteq S$ be picked at random from the set of subsets of size k of S. Given error bound ϵ and confidence parameter δ, if $k \geq \frac{\ln \frac{2}{\delta}}{2\epsilon^2}$, with probability at least $1 - \delta$, the ϕ-quantile of Q is an ϵ-approximation ϕ-quantile of S.*

Proof. The element with rank $\phi|Q|$ in Q[1] does not have rank within $(\phi \pm \epsilon)|S|$ in S if and only if one of the following holds: (a) More than $\phi|Q|$ elements in Q have rank less than $(\phi - \epsilon)|S|$ in S, or (b) more than $(1 - \phi)|Q|$ elements in Q have rank greater than $(\phi + \epsilon)|S|$ in S.

Since $|Q| = k$, the distribution of elements in Q are identical to the distribution where k elements are picked uniformly at random without replacement from S. This is due to the fact that any element of S is as likely to be included in Q as any other element in either scheme, and both schemes include k elements in Q.

Since the two distributions mentioned above are identical, we can think of the construction of Q as k random draws without replacement from a *0–1 box* that contains $|S|$ items, of which those with rank less than $(\phi - \epsilon)|S|$ are labelled "1" and the rest are labelled "0". For $i = 1, \ldots, k$, let X_i be the random variable for the label of the ith element in Q. Then $X = \sum_{i=1}^{k} X_i$ is the number of elements in Q that have rank less than $(\phi - \epsilon)|S|$ in S. Clearly, $E[X] = (\phi - \epsilon)k$. Hence, $Pr[X \geq \phi k] = Pr[X - E[X] \geq \phi k - (\phi - \epsilon)k] = Pr[X - E[X] \geq \epsilon k] = Pr[\frac{X}{k} - E[\frac{X}{k}] \geq \epsilon]$. This is at most $e^{-2\epsilon^2 k}$, by Hoeffding's Inequality. Note that Hoeffding's Inequality applies to random samples chosen without replacement from a finite population, as shown in Section 6 of Hoeffding's original paper [12], without the need for independence of the samples.

---

[1]Wherever rank in a set is mentioned, it should be understood that this rank is over a sequence obtained by sorting the elements of the set.

Similarly, it can be shown that the probability that more than $(1-\phi)|Q|$ elements in Q have ranked greater than $(\phi+\epsilon)|S|$ in S is also at most $e^{-2\epsilon^2 k}$. Setting $2e^{-2\epsilon^2 k} \leq \delta$, we have $k \geq \frac{\ln\frac{2}{\delta}}{2\epsilon^2}$. $\qquad\square$

We now show which layer we must use for given error and confidence bounds.

Theorem 3.2. *If a ϕ-quantile query is sent to layer $l < \log N - \log\left(\frac{\ln\frac{4}{\delta}}{2\epsilon^2} + \ln\frac{2}{\delta} + \sqrt{\ln\frac{2}{\delta}(2\frac{\ln\frac{4}{\delta}}{2\epsilon^2} + \ln\frac{2}{\delta})}\right)$, then the answer will be the ϵ-approximation ϕ-quantile of the whole network with probability greater than $(1 - \delta)$.*

Proof. By Lemma 3.1, the probability that layer $l < \log N - \log\left(k + \ln\frac{2}{\delta} + \sqrt{\ln\frac{2}{\delta}(2k + \ln\frac{2}{\delta})}\right)$ has fewer than k sensors is less than $\frac{\delta}{2}$. By Lemma 3.2, if the number of sensor nodes on layer l is at least $\frac{\ln 2(\frac{2}{\delta})}{2\epsilon^2} = \frac{\ln\frac{4}{\delta}}{2\epsilon^2}$, the probability that the ϕ-quantile on layer l is ϵ-approximation ϕ-quantile of the sensor network is at least $1 - \frac{\delta}{2}$. Hence, the answer returned by layer $l < \log N - \log\left(\frac{\ln\frac{4}{\delta}}{2\epsilon^2} + \ln\frac{2}{\delta} + \sqrt{\ln\frac{2}{\delta}(2\frac{\ln\frac{4}{\delta}}{2\epsilon^2} + \ln\frac{2}{\delta})}\right)$ is ϵ-approximation ϕ-quantile of the sensor network with probability greater than $(1 - \delta)$. $\qquad\square$

3.3.3 AVERAGE and SUM Queries

3.3.3.1 The Initial Algorithm

AVERAGE queries and SUM queries are correlated queries where the AVERAGE is just SUM$/N$. Since we know the number of the sensors in advance, we just analyze the AVERAGE queries in this section and do not explicitly explain the SUM queries.

We now consider approximating the average data value over the whole sensor network by querying a particular layer. The below lemma indicates that the expectation of the average data value of an arbitrary layer is the same as the average of the base layer, which is the exact average of the sensor network.

Lemma 3.3. *Let $a_1, a_2, \ldots, a_N$ be the data values collected by the nodes $s_1, s_2, \ldots, s_N$ of the sensor network. Let k be the number of sensors on layer l. Let $X_1, X_2, \ldots, X_k$ be the random variables describing the k data values on layer l. Let $\overline{X} = \frac{1}{k}\sum_{i=1}^{k} X_i$. Then $E[\overline{X}] = \frac{1}{N}\sum_{i=1}^{N} a_i$.*

Proof. Since each sensor independently promotes itself to layer l with the same probability, $Pr[X_i = a_1] = Pr[X_i = a_2] = \ldots = Pr[X_i = a_N] = \frac{1}{N}$, for $i = 1, 2, \cdots, k$. Then $E[X_i] = \frac{1}{N}(a_1 + a_2 + \cdots + a_N)$. Hence, $E[\overline{X}] = E[\frac{1}{k}\sum_{i=1}^{k} X_i] = \frac{1}{k}\sum_{i=1}^{k} E[X_i] = \frac{1}{k}\frac{k}{N}(a_1 + a_2 + \cdots + a_N) = \frac{1}{N}\sum_{i=1}^{N} a_i$. $\qquad\square$

Thus, we propose that the average returned by the queried layer be output as the average of the whole network. The next theorem shows that, given the appropriate layer, this constitutes an ϵ-approximation to the actual average with probability greater than $1 - \delta$.

Theorem 3.3. *Let the data value at each sensor come from the interval* $[a, b]$, *and let* l *be such that* $l < \log N - \log \left(\frac{(b-a)^2 \ln \frac{4}{\delta}}{2\epsilon^2} + \ln \frac{2}{\delta} + \sqrt{\ln \frac{2}{\delta}(2 \frac{(b-a)^2 \ln \frac{4}{\delta}}{2\epsilon^2} + \ln \frac{2}{\delta})} \right)$, *then the probability that the average of the data values on layer* l *deviates from the exact average by more than* ϵ *is less than* δ.

Proof. Let k be the number of sensors on layer l. As we have explained in Lemma 3.3, these k sensors can be considered to be random samples without replacement from all of the N sensors. Let $X_1, X_2, \ldots, X_k$ be the random variables describing the k sensor values on layer l, as in Lemma 3.3. Then $a \leq X_i \leq b$ for $i = 1, 2, \cdots, k$. Let $\overline{X} = \frac{1}{k} \sum_{i=1}^{k} X_i$. By Lemma 3.3, $E[\overline{X}]$ is the exact average of the sensor network. For any $\epsilon > 0$, $Pr[|\overline{X} - E[\overline{X}]| \geq \epsilon] \leq 2e^{\frac{-2k\epsilon^2}{(b-a)^2}}$, by Hoeffding's Inequality. Setting $2e^{\frac{-2k\epsilon^2}{(b-a)^2}} \leq \delta/2$, we have $k \geq \frac{(b-a)^2 \ln \frac{4}{\delta}}{2\epsilon^2}$. By Lemma 3.1, the probability that layer $l < \log N - \log \left(k + \ln \frac{2}{\delta} + \sqrt{\ln \frac{2}{\delta}(2k + \ln \frac{2}{\delta})} \right)$ has fewer than k sensors is less than $\frac{\delta}{2}$. Thus, if we send an AVERAGE Query to layer $l < \log N - \log \left(\frac{(b-a)^2 \ln \frac{4}{\delta}}{2\epsilon^2} + \ln \frac{2}{\delta} + \sqrt{\ln \frac{2}{\delta}(2 \frac{(b-a)^2 \ln \frac{4}{\delta}}{2\epsilon^2} + \ln \frac{2}{\delta})} \right)$, the probability that the estimated average deviates from the exact average more than ϵ is less than $\frac{\delta}{2} + \frac{\delta}{2} = \delta$.   □

3.3.3.2   Utilizing Statistical Information About the Behavior of Data

If we have access to additional information regarding the characteristics of the objects that the sensor network is monitoring, we can reduce the latency even further. In what follows, we show that the knowledge that the change in data values over time respects a certain distribution (such as the normal distribution) can be used to improve the quality of our estimates.

Assume the change of the data value in one time unit for each single sensor follows a normal distribution with mean μ. For instance, we might know that the electricity consumption is likely to rise around $10°$ from 2 a.m. to 12 p.m., and fall around $6°$ from 12 p.m. to 8 p.m. Small variations might happen but substantial changes are less likely (see Figs. 3.2 and 3.3). The change in the average value also follows a normal distribution since the sum of normal distributions is still a normal distribution with mean and variance equal to the sum of the individual means and variances.

Fig. 3.2 Electricity consumption changes from 2 a.m. to 12 p.m.

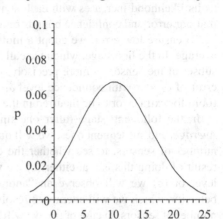

Temperature Change from 2am to 12pm

Fig. 3.3 Electricity consumption changes from 12 p.m. to 8 p.m.

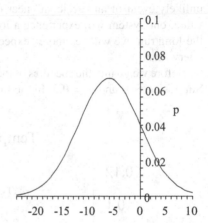

Temperature Change from 12pm to 8pm

To make use of the statistical information regarding the change in the value, we adopt a history-based approach, that we assume we know the distribution of the change of the environment to be a normal distribution with mean μ.

The intuition of our strategy is as follows. First we obtain an initial estimate avg of the average data value in the network, which, by our computations above, is likely to be close to the true average. After one unit of time, the true average is likely to have changed by some value close to μ. Thus, $avg + \mu$ is likely to be a good estimate for the average for that point in time. However, errors have been introduced into our estimate. One cause for possible error is the fact that only a subset of the sensors have been queried. The other contribution to the error comes from our inability to know the exact change in the data value; we only know from the normal distribution that the change is "likely" to be "around" μ. Since the quantity of the error, as well as its likelihood increases with each step of this procedure, we need to make sure that our error and confidence bounds remain at acceptable levels.

To ensure low error, we adopt a multi-stage approach to our estimation of the average. In the first stage, which we call *Query Average*, we query a relatively large subset of the sensors – more precisely, we query a low enough layer to obtain an error of $\epsilon_1 < \epsilon$ with confidence level $\delta_1 < \delta$. This high guarantee will leave some room for extra error to be incurred in the following stages.

In the following stage (after one time unit has elapsed), which we call *Test Average*, and subsequent ones, we will query higher layers, thus involving a smaller number of sensors, to see whether the expected change pattern is followed. The result of doing this is that either (a) we will boost the confidence to an acceptable level or (b) we will observe an "anomaly", that is, a deviation from expected behavior, which we will attempt to resolve by querying a lower layer with a larger number of sensors. In case of (a), we will have obtained a fast and acceptable answer by querying only a very small number of sensors. Case (b) on the other hand is, by definition of the normal distribution, an anomaly that will not often happen. In the unlikely event of an "accident" near one of the nodes, in the form of an atypical value, our system will experience a longer query time for the sake of accuracy. In the long run, we will see more "expected" cases and will observe a lower average query time.

Before we go into the specifics of the algorithm, we present an example (Fig. 3.4). Suppose $\epsilon = 8$ and $\delta = 0.2$. In the first stage, we see that we get the average data

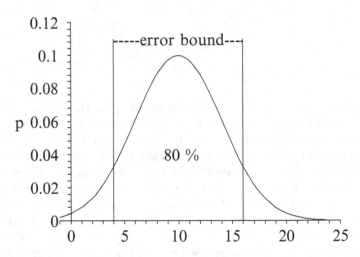

Fig. 3.4 The possible change for electricity consumption after a time unit follows a normal distribution with $\mu = 10$ and $\sigma = 4$. To ensure the ultimate error bound of $\epsilon = 8$, the error bound $\epsilon_n = 6$

value $avg_1 = 60\,°F$ by using error bound, say $\epsilon_1 = 2$ and a confidence level, say 0.9, (i.e., error probability $\delta_1 = 0.1$). After 10 h, we expect that the electricity consumption changes to $avg_1 + \mu = 70\,°F$. Let $\hat{\epsilon}$ and $\hat{\delta}$ denote the confidence interval and confident level of the normal distribution. To ensure an error within the user specified bound of 8, then $\hat{\epsilon} \leq 6$, as shown in Fig. 3.4, resulting a confident level of $\hat{\delta} = 0.8$. Therefore, after this period of time, the overall confidence level is $0.8 \times 0.9 = 0.72$, i.e., the error probability is larger than the 0.2 specified as acceptable. To boost the confidence level to 0.8, we need to query a few more sensors with an error bound of $\epsilon_2 = 8$ and a much looser error probability of $\delta_2 = \frac{0.2}{1-0.72} = 0.714$. If the returned value of Test Average is 70, we will return this value. Otherwise, if the returned value falls outside of 70 ± 8, this indicates that an anomaly might be present,[2] in which case we perform Query Average to determine the new electricity consumption value. If the returned value falls within 70 ± 8, to ensure the error bound, we perform Test Average with a more stringent error bound until an anomaly is found or the new average value is confirmed.

Below we explain our algorithm in higher detail and analyze its properties mathematically. Figure 3.5 shows Algorithm Query Average. It takes as input the error and confidence parameters ϵ, δ. We assume that Query(l) returns the average data value for sensors on layer l.

Our next algorithm (Fig. 3.6) shows how to perform Test Average given Query Average. It takes as input the error and confidence parameters ϵ, δ, as well as the mean μ and standard deviation σ of the distribution of the change of the data value. Also, it takes avg_1 which is the average obtained from Query Average and the round number i.

In Line 2 of Algorithm Test Average, we calculate the probability that the change will fall within interval $\hat{\epsilon}$. In Lines 1–5, if the Query Average has already guaranteed the error probability, we do not perform any further queries. This might occur when the number of sensors queried in Query Average is large enough. Line 12 displays the threshold that we should perform the query again.

Theorem 3.4. *Assume the data value collected by each sensor is bounded by* $[a, b]$ *and the change in the average of the values at all sensors follows a normal*

Algorithm QueryAvg (ε, δ)
1 Select $\varepsilon_1 < \varepsilon$ and $\delta_1 < \delta$.
2 $l_1 = \log N - \log\left(\frac{(b-a)^2 \ln\frac{4}{\delta_1}}{2\varepsilon_1^2} + \ln\frac{2}{\delta_1} + \sqrt{\ln\frac{2}{\delta_1}\left(2\frac{(b-a)^2\ln\frac{4}{\delta_1}}{2\varepsilon_1^2} + \ln\frac{2}{\delta_1}\right)} \right)$
3 $avg_1 =$ Query(l_1).
4 **return** avg_1

Fig. 3.5 Algorithm Query Average

---

[2] Here we use the word anomaly to indicate a situation whose likelihood is small according to the given normal distribution.

Algorithm TestAvg $(i, \varepsilon, \delta, \mu, \sigma, avg_1, \varepsilon_1, \delta_1)$

1 $\hat{\varepsilon} = \varepsilon - \varepsilon_1$.

2 Calculate $\hat{\delta} = Pr(\mu - \varepsilon_n < X < \mu + \varepsilon_n)$ by Normal Distribution.

3 **if** $1 - \hat{\delta} \times (1 - \delta_1) < \delta$,

4    $avg_i = avg_1 + \mu$, **return** avg_i

5 **else**

6    $k = 0, \delta_i = \frac{\delta}{1 - \hat{\delta} \times (1 - \delta_1)}$

   repeat

7    $\varepsilon_k = \varepsilon - k, \varepsilon_i = \varepsilon_k$

8    $l_i = \log N - \log \left(\frac{(b-a)^2 \ln \frac{4}{\delta_i}}{2\varepsilon_i^2} + \ln \frac{2}{\delta_i} + \sqrt{\ln \frac{2}{\delta_i} (2 \frac{(b-a)^2 \ln \frac{4}{\delta_i}}{2\varepsilon_i^2} + \ln \frac{2}{\delta_i})} \right)$

9   **if** $(\text{Query}(l_i) \leq avg_1 + \mu + k)$
         and $(\text{Query}(l_i) \geq avg_1 + \mu - k)$

10   **then** $avg_i = avg_1 + \mu$, **return** avg_i

11   **else** increase k

12   **if** $k \geq \varepsilon$, Goto QueryAvg

Fig. 3.6 Algorithm Test Average

distribution with mean μ and standard deviation σ. The probability that algorithm QueryAvg and TestAvg will deviate from the exact average by more than ϵ is less than δ.

Proof. Let us first consider QueryAvg. Choose any ϵ_1, δ_1 such that $\epsilon_1 < \epsilon$ and $\delta_1 < \delta$. By Theorem 3.3, we can obtain the desired accuracy by sending queries to any layer l_1 where $l_1 < \log N - \log \left(\frac{(b-a)^2 \ln \frac{4}{\delta_1}}{2\epsilon_1^2} + \ln \frac{2}{\delta_1} + \sqrt{\ln \frac{2}{\delta_1} (2 \frac{(b-a)^2 \ln \frac{4}{\delta_1}}{2\epsilon_1^2} + \ln \frac{2}{\delta_1})} \right)$.

For TestAvg, let $\overline{Y}_i$ denote the average value over all the sensors at round i. Define $\Delta_i = \overline{Y}_i - \overline{Y}_1$. We know *a priori* that the probability distribution for each Δ_i is a normal distribution with mean μ_i and variance σ_i^2.

In round i, $\hat{\varepsilon} = \epsilon - \epsilon_1$ is the confidence interval for normal distribution, hence, $\hat{\delta}$ is probability that the change in the value of the average will not exceed $\hat{\varepsilon}$.

Therefore, with probability $1 - \hat{\delta} \times (1 - \delta_1)$, we can guarantee an error bound of $\epsilon_1 + \hat{\varepsilon} < \epsilon$. If $1 - \hat{\delta} \times (1 - \delta_1) < \delta$, then our query satisfies both bounds ϵ and δ, and we can compute the value to be returned from the value in the previous round and the expected change. Otherwise, we choose $\delta_i = \frac{\delta}{1 - \hat{\delta} \times (1 - \delta_1)}$ which ensures that the confidence error δ will be bounded in the ith round. For error bound ϵ_i in the ith round, since we do not know the returned value, we can use all $\epsilon_i = \epsilon_k$ where $\epsilon \geq \epsilon_k \geq 0$ as long as the returned value $avg_i \pm (\epsilon_k)$ will be bounded by $(avg_1 + \mu) \pm \epsilon$. If that happens, the change is confirmed. Otherwise, to bound ϵ and δ, a new QueryAvg must be performed. In our algorithm, we reduce ϵ_i iteratively from ϵ to 0, and use ϵ_i and δ_i to query layer $l_i < \log N -$ $\log \left(\frac{(b-a)^2 \ln \frac{4}{\delta_i}}{2\epsilon_i^2} + \ln \frac{2}{\delta_i} + \sqrt{\ln \frac{2}{\delta_i} (2 \frac{(b-a)^2 \ln \frac{4}{\delta_i}}{2\epsilon_i^2} + \ln \frac{2}{\delta_i})} \right)$ so that the number of sensors in TestAvg will increase little by little and stop as early as possible. $\square$

3.3.4 Effect of the Promotion Probability p

The promotion probability p will only affect the logarithmic base of the system. Therefore, we only give out comparable theorems without detailing the proofs.

Theorem 3.5 (w.r.t. Theorem 3.2). *To attain the ϕ-quantile of the sensor readings with error bound ϵ and confidence level δ, the query must be sent to layer $l <$*
$$\log_{\frac{1}{p}} N - \log_{\frac{1}{p}} \left(\frac{ln\frac{4}{\delta}}{2\epsilon^2} + ln\frac{2}{\delta} + \sqrt{ln\frac{2}{\delta}(2\frac{ln\frac{4}{\delta}}{2\epsilon^2} + ln\frac{2}{\delta})} \right), \text{ then the answer will be the } \epsilon\text{-}$$
approximation ϕ-quantile of the whole network with probability greater than $(1-\delta)$.

Theorem 3.6 (w.r.t. Theorem 3.3). *Let the data value at each sensor come from the interval $[a, b]$, and let l be such that $l < \log_{\frac{1}{p}} N - \log_{\frac{1}{p}} \left(\frac{(b-a)^2 ln\frac{4}{\delta}}{2\epsilon^2} + ln\frac{2}{\delta} + \right.$*
$$\left. \sqrt{ln\frac{2}{\delta}(2\frac{(b-a)^2 ln\frac{4}{\delta}}{2\epsilon^2} + ln\frac{2}{\delta})} \right), \text{ then the probability that the average of the data values}$$
on layer l deviates from the exact average by more than ϵ is less than δ.

The algorithms of QueryAvg and TestAvg can be adjusted accordingly and the corresponding theorems also follow.

3.4 Energy Consumption

It can be readily observed that in our system higher layer sensors will be transmitted at longer ranges than their lower layer counterparts. Given that any high layer sensor is also presented in all the lower layers, if nothing is done to balance out the energy consumption, the higher layer sensors may get depleted much faster than the lower layer ones. To balance out the energy consumption, our system reconstructs the layered network periodically by deciding each layer from scratch, so that the top layer sensors change over time. An appropriate timing scheme for the reconstruction will lead to relatively uniform energy consumption across the sensors in the network. Note that the frequency of reconstructions has no expected effect on accuracy, since we are as likely to be stuck with a "good" sample of sensors (in which case reconstruction is likely to give us a worse sample) as with a "bad" one. Given the above and the overhead of building a new aggregation tree for each new construction, we are interested in infrequently repeating this procedure for making the energy consumption more even across sensors.

Let the lifetime of the network be the time between its initial construction and the first time that a sensor runs out of power [24]. We investigate the trade-off between the timing of the reconstructions and the expected lifetime of our system in our simulations. In this section, we analyze our system assuming that each sensor has sufficient power to let it undergo several reconstructions, and that we run reconstructions enough times. Ideally, we have a totally symmetric scenario where the service that each sensor has performed on each layer is identical across sensors. Since the layers are chosen in a randomized fashion, given a large enough number of reconstructions that one will see that most sensors have served on different layers.

The energy spent by each sensor for a query directly depends on the distance between the sensor and its neighbors. Recall that since there are an expected $(N/2^l)$ nodes on layer l, the transmission range is set to be $r(l) = \frac{D}{\sqrt{N/2^l}}$. Therefore, the energy spent by each sensor for each query on layer l is $e(l) = \left(\frac{D}{\sqrt{N/2^l}}\right)^\alpha$, which is that we will use below to estimate the overall system lifetime.

3.4.1 Overall Lifetime of the System

In this section, we assume that the queries are uniformly distributed across different layers due to the error bounds and confidence levels coming independently from the users.

We now present a theorem which estimates the expected lifetime of our system depending on the network parameters.

Theorem 3.7. *In a setting where each level is equally likely to be queried, the expected lifetime of our system is* $E(t) = \frac{BL(\sqrt{N})^\alpha(1-(\sqrt{2})^{\alpha-2})}{\lambda D^\alpha(1-(\sqrt{2})^{L(\alpha-2)})}$

Proof. We assume that each layer has the same probability $\frac{1}{L}$ of being queried. The probability that a sensor exists on layer l is $\frac{1}{2^l}$, therefore, the energy consumption for this sensor is $\sum_{l=0}^{L-1} \frac{1}{2^l} e(l) \frac{1}{L}$. Let the life expectancy of this sensor be t. Recall that B is the battery power, λ is the incoming query interval following (assumed) Poisson distribution, and $e(l) = \left(\frac{D}{\sqrt{N/2^l}}\right)^\alpha$. We have $\sum_{l=0}^{L-1} \frac{1}{2^l} e(l) \frac{1}{L} \lambda t = B$. Therefore, the expected lifetime of the system is $E(t) = \frac{BL(\sqrt{N})^\alpha(1-(\sqrt{2})^{\alpha-2})}{\lambda D^\alpha(1-(\sqrt{2})^{L(\alpha-2)})}$ □

3.5 Evaluation Results

We use numerical simulations to test the performance of our system, as well as to observe the effects of the parameters of the algorithm and the re-election time on the performance.

3.5.1   System Settings

We set the default number of sensors to be $N = 10,000$. The default promotion is probability $p = \frac{1}{2}$. We focus on QUANTILE and AVERAGE queries in our simulations.

3.5.2   Layers vs. Accuracy

We first evaluate the trade-off between the layer answering a query and parameters relating to the quality of the answer to the query.

3.5.2.1   QUANTILE Queries

We first study QUANTILE Queries in Fig. 3.7. It can be seen clearly that, as ϵ and δ increase, the layer that the query should be sent to also increases, as confirmed by our computations. Here it can be observed that even though the layer number monotonically increases with both parameters, ϵ has more impact on it than δ as can be seen from Fig. 3.7a, b. This is because the confidence parameter δ can easily be improved using standard boosting techniques from probability theory and randomized algorithms. In fact, repeating the algorithm $O(\log k)$ times and returning the median answer will improve δ to δ/k, since the probability of getting an incorrect answer $(\log k)/2$ times is at most $\delta^{\log k}$, which is $O(\delta/k)$. On the other hand, to reduce ϵ by a constant factor k, $O(k)$ repetitions of the experiment is needed. Figure 3.7c shows the trade-off between p and the layer number. As p increases, the variation in the layer number for the same query is more obvious. This is because there are fewer layers for smaller p and the choice of layer is more coarse-grained than for larger p.

3.5.2.2   AVERAGE Queries

We show the trade-off of the layer number with δ, ϵ and p for average queries in Fig. 3.8. We observe the same effect as with the QUANTILE queries, i.e., ϵ has a larger impact than δ, for the same reason. This gives us hints for building test queries as we have additional statistical information.

To investigate the question of how to choose the parameters introduced in our algorithms, QueryAvg and TestAvg, we fix the following parameters and vary the others. The upper and lower bounds of the data values are $a = 20$ and $b = 100$. We set the user-defined error bound and confidence parameter to be $\epsilon = 15.1$ and $\delta = 25\%$ respectively. The mean and standard deviation for the normal distribution are set to $\mu = 20$ and $\sigma = 8$. We choose ϵ_1 in a range of $[6, 15]$, and two different

Fig. 3.7 QUANTILE
Queries, the relation among
the layer as a function of the
confident parameter δ, the
error parameter ϵ and the
promotion probability p.
(**a**) Layer as a function of δ.
(**b**) Layer as a function of ϵ.
(**c**) Layer as a function of p

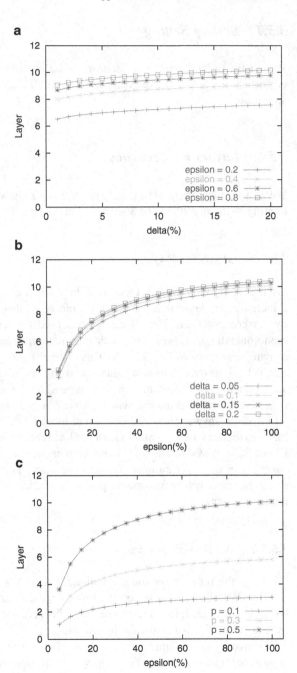

Fig. 3.8 AVERAGE Queries, the relation among the layer as a function of the confident parameter δ, the error parameter ϵ and the promotion probability p.
(a) Layer as a function of δ.
(b) Layer as a function of ϵ.
(c) Layer as a function of p

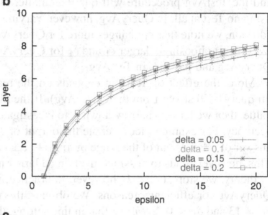

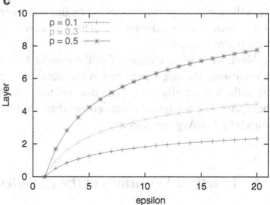

δ_1, 0.05 and 0.2 respectively, to compare this two-phase algorithm with no Test at all, i.e., using ϵ and δ directly.

Note that, when $p = 0.5$, the expected number of sensors in each layer increases by 2 as we go down each layer. To reduce overall delay, every query performs on some layer $i - 1$ rather than layer i must be compensated by 2 or more runs of TestAvg performed on layer $i + 1$ rather than layer i, or 1 or more on $i + 2$ rather than $i + 1$, etc. Thus, the combination of QueryAvg and TestAvg is more profitable when the change in the data is highly predictable. Thus, QueryAvg and TestAvg can be used for emergency monitoring applications in stable environments whereas QueryAvg alone can be used in applications of data acquisition in changing environments.

Figure 3.9a shows the effect of ϵ_1 and δ_1 for QueryAvg. Figure 3.9b shows the effect of ϵ_1 and δ_1 for TestAvg. In our simulations, we see that regardless of the choice of ϵ_1 and δ_1, the QueryAvg procedure will query a larger number of sensors and the TestAvg procedure will query a smaller number of sensors, comparing to using no Test at all. In QueryAvg, however, varying δ_1 has relatively larger effect. In addition, we note that ϵ_1 changes more for QueryAvg than for TestAvg. As a result, we suggest choosing a larger ϵ_1 and δ_1 for QueryAvg, so that we can save more for QueryAvg and pay less in TestAvg.

More, the effect of TestAvg depends on the readings of the second phase, (or ith query of TestAvg from the QueryAvg). If the ith query is far from the expected value, then we have to narrow down ϵ_i to investigate the sensor network with higher accuracy. For example, let r denote the output of TestAvg, in our algorithm, if the answer of $r \pm \epsilon_i$ is out of the range of $avg_1 + \mu \pm \epsilon$, then we set ϵ_i more stringently. This, however, leads to possibly involving a larger number of sensors for this query. In theory, we stop at $\epsilon_i = 0$, however, in practice, we can stop earlier and move to QueryAvg for efficiency reasons. We observe this effect in Fig. 3.9c where we set $\epsilon_1 = 13$ and $\delta_1 = 0.2$. We see that in this setting, there are reasonable chances that we will stop with savings since when the results from TestAvg fall within $avg_1 + \mu \pm 5$, fewer sensors are queried. This also confirms that ϵ has a greater impact than δ for our architecture.

Next, we study the effect of σ, the standard deviation of the normal distribution. σ represents the rate of change in the data. One can see in Fig. 3.10 that σ has a big influence on efficiency. The discontinuity of the lines in Fig. 3.10a, b indicate that QueryAvg has tested more sensors than required, making some of the following rounds of TestAvg unnecessary.

3.6   Practical Variations of the Architecture

We have given the theoretical analysis of a layer architecture. We now discuss some practical concerns in this section.

Fig. 3.9 AVERAGE Queries with different δ_1 values; no Test denotes using the original δ and ϵ. (**a**) Query Average; layer as a function of ϵ_1. (**b**) Test Average; layer as a function of ϵ_1. (**c**) Test Average based on different readings; layer as a function of ϵ_i

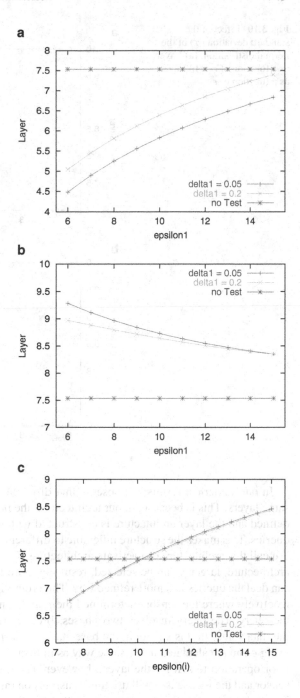

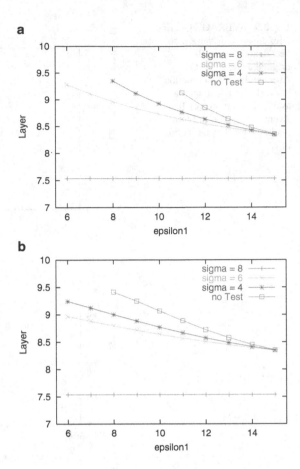

Fig. 3.10 Effect of the standard deviation (σ) of the normal distribution. (**a**) Layer as a function of ϵ_1. (**b**) Layer as a function of ϵ_1

In our numerical results, we observe that different ϵ and δ values may match to same layers. This is because, in our architecture, the promotion probability p is pre-defined and the layer architecture is constructed with no consideration for different queries. This makes the structure inflexible to different user parameters, e.g., ϵ and δ. We call this architecture pro-activate architecture. To solve this problem within this architecture, larger p can be selected, resulting more layers. Thus, the architecture can deal the queries in a more refined way. The promotion may also have determined reactively where the sensors determined the p and construct a layer after receiving the query. The query involves two phases, broadcasting the query and collecting the data value. In this situation, the base station determines the transmission range using ϵ and δ, calculated in the same way as in Sect. 3.3. We do not need to take the floor operation to suit for the layers, however. The transmission is sent to the root sensor and the root sensor will use this transmission range to broadcast the message. Every sensor receiving this message will relay this information using the same transmission range. Notice that by using this larger transmission range (compared to the base layer range), the delay of the broadcasting phase will be controlled.

All sensors, after receiving this query, will calculate the promotion probability p which is inversely proportion to the transmission range. Sensors that succeed the promotion will then send the data value back to the root.

In the above settings, all nodes have to relay the message using a large transmission range in the broadcasting session. The energy consumption is thus much higher than the pro-activate architecture. Several optimizations can be applied for saving energy consumption. For example, since when a sensor sends a message, all neighbor sensors within the transmission range will hear this message [2], therefore, only the neighbors that are "farther" away to this sensor will relay the message. We do not elaborate on the broadcasting and collecting techniques as these are not the focus of this chapter. Detailed studies in this area can be found in [13, 18, 24, 25].

3.7 Summary and Discussions

In this chapter, we present one problem related to wireless sensor data collection. We face a trade-off between delay sensitive requirement and data accuracy. We observe that the delay issue can be translated into reducing the number of data collected. This is where the sublinear algorithm jumps in. Such linkage can be common in applying sublinear algorithms to real-world applications.

From sublinear algorithm point of view, we see how we jointly use multiple sublinear algorithms to achieve a certain task, i.e., which layer a query should send to. In addition, we use side information to further enhance the sublinear algorithms. Meanwhile, we note that when facing a real problem, things go beyond algorithm development. In this chapter, we have created a layered architecture for the wireless sensor networks and discussed in details about how the layered architecture should be constructed and how the queries should be handled. What's more, we discuss how the layered structure should be periodically re-constructed for energy load balancing.

References

1. I. Akyildiz, W. Su, Y. Sankarasubramaniam, and E. Cayirci, "A Survey on Sensor Networks", in *IEEE Communications Magazine*, vol. 40, no. 8, pp. 102–114, 2002.
2. J. Al-Karaki and A. Kamal, "Routing Techniques in Wireless Sensor Networks: A Survey", in *IEEE Wireless Communications*, vol. 11, no. 6, pp. 6–28, 2004.
3. S. Banerjee and S. Khuller, "A Clustering Scheme for Hierarchical Control in Multi-hop Wireless Networks", in *Proc. IEEE INFOCOM'01*, Anchorage, AK, Apr. 2001.
4. A. Boukerche, R. Pazzi, and R. Araujo, "A Fast and Reliable Protocol for Wireless Sensor Networks in Critical Conditions Monitoring Applications", in *Proc. ACM MSWiM'04*, Venice, Italy, Oct. 2004.

5. D. Braginsky and D. Estrin, "Rumor Routing Algorithm for Sensor Networks", in *Proc. WSNA'02*, Atlanta, GA, Sep. 2002.
6. C. Buragohain, D. Agrawal, and S. Suri, "Power Aware Routing for Sensor Databases", in *Proc. IEEE INFOCOM'05*, Miami, FL, Mar. 2005.
7. M. Chu, H. Haussecker, and F. Zhao, "Scalable Information-Driven Sensor Querying and Routing for Ad Hoc Heterogeneous Sensor Networks", in *International Journal of High Performance Computing Applications*, vol. 16, no. 3, pp. 293–313, 2002.
8. D. Estrin, R. Govindan, J. Heidemann, and S. Kumar, "Next Century Challenges: Scalable Coordination in Sensor Networks", in *Proc. ACM MOBICOM'99*, Seattle, WA, Aug. 1999.
9. M. Greenwald and S. Khanna, "Power-Conserving Computation of Order-Statistics over Sensor Networks", in *Proc. ACM PODS'04*, Paris, France, Jun. 2004.
10. W. Heinzelman, J. Kulik, and H. Balakrishnan, "Adaptive Protocols for Information Dissemination in Wireless Sensor Networks", in *Proc. ACM MOBICOM'99*, Seattle, WA, Aug. 1999.
11. W. Heinzelman, A. Chandrakasan, and H. Balakrishnan, "Energy-Efficient Communication Protocol for Wireless Microsensor Networks", in *Proc. Hawaaian International Conference on Systems Science (HICSS'00)*, Wailea Maui, HI, Jan. 2000.
12. W. Hoeffding, "Probability Inequalities for Sums of Bounded Random Variables", in *Journal of the American Statistical Association*, vol. 58, no. 301, pp. 13–30, 1963.
13. C. Intanagonwiwat, R. Govindan, and D. Estrin, "Directed Diffusion: A Scalable and Robust Communication Paradigm for Sensor Networks", in *Proc. ACM MOBICOM'00*, Boston, MA, Aug. 2000.
14. C. Intanagonwiwat, D. Estrin, R. Govindan, and J. Heidemann, "Impact of Network Density on Data Aggregation in Wireless Sensor Networks", in *Proc. IEEE ICDCS'02*, Vienna, Austria, Jul. 2002.
15. V. Kawadia and P. Kumar, in *the Power Control and Clustering in Ad-hoc Networks*, in *Proc. IEEE INFOCOM'03*, San Francisco, CA, Mar. 2003.
16. B. Krishnamachari, D. Estrin, and S. Wicker, "the Impact of Data Aggregation in Wireless Sensor Networks", in *Proc. IEEE ICDCS Workshop on Distributed Event-based System (DEBS'02)*, Vienna, Austria, Jul. 2002.
17. S. Lindsey and C. Raghavendra, "PEGASIS: Power-Efficient Gathering in Sensor Networks", in *IEEE Aerospace Conference Proceedings*, vol. 3, pp. 1125–1130, 2002.
18. S. Madden, R. Szewczyk, M. Franklin, and W. Hong, "Supporting Aggregate Queries over Ad-Hoc Wireless Sensor Networks", in *Proc. IEEE International Workshop on Mobile Computing Systems and Application (WMCSA'02)*, Callicon, NY, Jun. 2002.
19. S. Madden, M. Franklin, J. Hellerstein, and W. Hong, "Tag: A Tiny Aggregation Service for Ad hoc Sensor Networks", in *Proc. USENIX OSDI'02*, Boston, MA, Dec. 2002.
20. N. Sadagopan, B. Krishnamachari, and A. Helmy, "the ACQUIRE Mechanism for Efficient Querying in Sensor Networks", in *Proc. the IEEE International Workshop on Sensor Network Protocol and Applications (SNPA'03)*, Seattle, WA, May. 2003.
21. N. Shrivastava, C. Buragohain, D. Agrawal, and S. Suri, "Medians and Beyond: New Aggregation Techniques for Sensor Networks", in *Proc. ACM SENSYS'04*, Baltimore, MD, Nov. 2004.
22. J. Wieselthier, G. Nguyen, and A. Ephremides, "On the Construction of Energy-Efficient Broadcast and Multicast Trees in Wireless Networks", in *Proc. IEEE INFOCOM'00*, Tel-Aviv, Israel, Mar. 2000.
23. Y. Yao and J. Gehrke, "Query Processing for Sensor Networks", in *Proc. CIDR'03*, Asilomar, CA, Jan. 2003.
24. O. Younis and S. Fahmy, "Distributed Clustering in Ad-hoc Sensor Networks: A Hybrid, Energy-Efficient Approach", in *Proc. IEEE INFOCOM'04*, Hong Kong, China, Mar. 2004.
25. Y. Yu, B. Krishnamachari, and V. Prasanna, "Energy-Latency Tradeoffs for Data Gathering in Wireless Sensor Networks", in *Proc. IEEE INFOCOM'04*, Hong Kong, China, Mar. 2004.

Chapter 4
Application on Big Data Processing

4.1 Introduction

4.1.1 Big Data Processing

Today's lightening-fast generation of data from massive sources and advanced data analytics have made mining the information from *big data* possible. We have witnessed the success of many big data applications. For example, Amazon uses its massive historical shipment tracking data to recommend goods to targeted customers, and Google uses billions of query data to predict flu trends, which can sometimes do one week earlier than the National Centers for Disease Control and Prevention (CDC).

The processing of big data, however, imposes unprecedented demands on the underlying computing and networking infrastructures. For input data at the petabyte or even exabyte scale, simply improving the power of individual machines, that is *scale-up*, is hardly practical. State-of-the-art tools, most notably MapReduce, are generally performed on dedicated server clusters to explore data parallelism. They rely on large-scale machines that work together, that is, *scale-out*, to process the big data in a divide-and-conquer manner.

As compared to conventional data processing, tools resembling MapReduce are still new to the market and there is much room for improvement. There have been many studies on improving the performance of MapReduce. In this chapter, we specifically study a key inefficiency in the current MapReduce framework when facing skewed distributions of data.

© The Author(s) 2015
D. Wang, Z. Han, *Sublinear Algorithms for Big Data Applications*,
SpringerBriefs in Computer Science, DOI 10.1007/978-3-319-20448-2_4

4.1.2 Overview of MapReduce

We first present some background on the MapReduce framework. Among the many tools that scale-out the processing of big data to parallel machines, MapReduce (proposed by Google), is arguably the most popular, and has become the *de facto* standard. Figure 4.1 illustrates the basic MapReduce structure. A MapReduce job consists of two phases (or tasks), namely, *map* and *reduce*. Accordingly, there are two program functions: *mapper* and *reducer*. In the map phase, the input data is split into blocks. The mappers then scan the data blocks and produce intermediate data. The reduce phase starts from a *shuffle* sub-phase, which, run by the reducers, shuffles intermediate data and moves them to the corresponding nodes for reducing. The reducers then process the shuffled data and generate the final results. For complex problems, multiple rounds of map and reduce can be executed.

Its generic and simple interface also implies that MapReduce can be a bottleneck in the overall processing with specific applications or specific data. Significant efforts have been made on relieving blocking operations [15], improving energy efficiency [13, 14], enhancing scheduling [18, 26], or relaxing the single fixed dataflow [2, 6, 17, 25]. There have also been recent works on the efficient scheduling of massive MapReduce jobs running in parallel [3, 4, 22].

4.1.3 The Data Skew Problem

Most of the studies have assumed that the input data are of uniform distribution, which, often being hashed to reduce worker nodes, naturally leads to a desirable balanced load in the later stages. The real-world data, however, are not necessarily

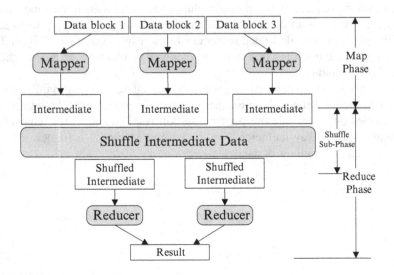

Fig. 4.1 The MapReduce architecture

uniform, and often exhibit a remarkable degree of skewness. For example, in PageRank, the graph commonly includes nodes with much higher degrees of incoming edges than others [11], and in Inverted Index, certain content can appear in many more documents than others [12]. Such a skewed distribution of the input or intermediate data can result in a small number of mappers or reducers taking a significantly longer time to complete than others [16]. Recent experimental studies [11, 12, 16] have shown that, in the CloudBurst application with a biology dataset of a bimodal distribution [20], the slowest map task takes five times as long to complete as the fastest. PageRank with Cloud9 data [24] is even worse, as the slowest map task takes twice as long to complete as the second slowest, and the latter is already five times slower than the average. Our experiments with the WordCount application produced similar results. Given that the overall finishing time is bounded by the slowest task, it can be dramatically prolonged with such skewed data.

In distributed databases, data skew is a known common phenomenon. There are mature solutions such as joining, grouping, aggregation, and others [21, 23]. Unfortunately, these can hardly be applied in the MapReduce context. The map function transfers the raw input data into (key, value) pairs, and the reduce function merges all intermediate values associated with the same intermediate key. In the database case, the pairs that share the same key do not need to be processed in a single machine. MapReduce, on the other hand, must guarantee that these pairs belong to the same partition—in other words, that they are distributed to the same reducer.

There are pioneering works dealing with the data skew in MapReduce [9, 12, 19]. Most of them are on offline heuristics, where the solution is to wait for all of the mappers to finish so as to obtain the key frequencies, or to engage in sampling before the map tasks to estimate the data distribution and then partition in advance, or to repartition the reduce tasks to balance the load among the servers. These approaches can be time-consuming with excessive I/O costs or network overheads. The solutions also lack theoretical bounds, given that most of them are heuristics.

4.1.4 Chapter Outline

In this chapter we, for the first time, examine the problem of accommodating data skew in MapReduce with online operations. In contrast with earlier solutions that address the problem in the very late reduce stage [12] or after seeing all of the data [8], we address the skew from the very beginning of the inputting of data, and make no assumptions about *a priori* knowledge of the distribution of the data, nor require synchronized operations. We examine the keys in a continuous fashion and adaptively assign the tasks with a load-balanced strategy. We show that the optimal strategy is a constrained version of the *online minimum makespan problem* [7], and demonstrate that, in the MapReduce context where tasks with identical keys must be scheduled to the same machine, there is an online algorithm with a provable 2-competitive ratio. We further suggest that the online solution can be enhanced by a sample-based algorithm, which identifies the most frequent keys and assigns associated tasks in advance. We show that, probabilistically, it achieves a 3/2-competitive ratio with a bounded error.

Note that in the development of our algorithm, there is an embedded sublinear algorithm. More specifically, our first online algorithm provides a baseline competitive ratio where there is no requirement to "peek" at the data in advance. Clearly, the ratio provided by this algorithm is loose. To tighten such a competitive ratio, we "peek" at a sample set of data and develop an advanced algorithm based on this information. Intuitively, such information helps us to obtain some knowledge of which data are more important, so that when handling the skewness of these data, it is possible to apply a more refined scheme. We develop the algorithm, study how much data (sublinear to the whole data set) we should "peek" at, and analyze a tightened competitive ratio.

We evaluate our algorithms on both synthetic data and a real public data set. Our simulation results show that, in practice, the maximum loads of our online and sample-based algorithms are close to those of the offline solutions, and are significantly lower than those with the naive hash function in MapReduce. They enjoy comparable computation times as those with the hash function, which are much shorter than those of the offline solutions.

In Sect. 4.2, we analyze in detail the data skew problem in MapReduce. We then formulate the problem. In Sect. 4.3, we present the first online algorithm, which achieves a baseline 2-competitive ratio. We further develop our enhanced online algorithm in Sect. 4.4, which has a sublinear algorithm embedded within. We evaluate the performance of our algorithms in Sect. 4.5. In Sect. 4.6, we further summarize this application and discuss how the sublinear algorithms are used in this application.

4.2   Server Load Balancing: Analysis and Problem Formulation

4.2.1   Background and Motivation

The MapReduce libraries have been written in different programming languages. We take Apache Hadoop (high-availability distributed object-oriented platform), one of the most popular free implementations, as an example. Hadoop is based on a master-worker architecture, where a master node makes scheduling decisions and multiple worker nodes run tasks dispatched from the master.

In the map phase, the master node divides the large dataset into small blocks and distributes them to the map workers. The map workers generate a large amount of intermediate (key, value) pairs and report the locations of these pairs on the local disk to the master, who is responsible for forwarding these locations to the reduce workers.

A hash function then assigns the values to different worker nodes for processing in the reduce phase. In Hadoop, the default hash function is

Hash(HashCode(intermediate key) mod ReducerNumber)

This is a simple hash function. It is highly efficient and naturally achieves load balance if the keys are uniformly distributed. This, however, can fail with skewed inputs. We look at an example of WordCount. WordCount is a classic MapReduce application for counting the number of words in a big document. It is also a commonly used benchmark application to evaluate the performance of MapReduce whenever an improvement is developed.

In a document, popular words such as "the," "a," and "of" appear much more frequently than other words. After hashing, they impose heavier workloads on the corresponding reduce workers. Consider a toy example shown in Fig. 4.2 with a skewed input. The naive hash function will assign keys *a*, *d*, and *g* to the first machine, keys *b* and *e* to the second machine, and keys *c* and *f* to the third machine. As a result, the first machine will achieve a maximum load with 19, six times more than that of the least load, while the maximum load of the balanced solution will be 12, as shown in the "optimal" row. Since the overall finishing time is bounded by the slowest, such simple hash-based scheduling is simply not satisfactory.

It is also worth noting that if an algorithm is designed to handle such load balancing problem, the algorithm must be online. This is because Hadoop starts to execute the reduce phase before every corresponding partition is available, i.e., it is activated when only part of the map phase has been completed (5 % by default) [12]. The rationale behind this synchronous operation is to overlap the map and the reduce phases and consequently reduce the maximum finishing time; yet it can prevent making a partition of the map phase and the reduce phase in advance. In fact, the

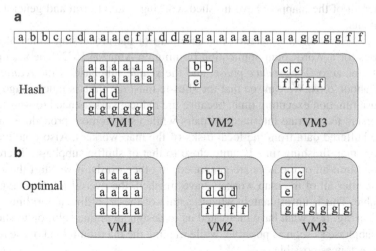

Fig. 4.2 An illustrative example for key assignment in MapReduce. There are three machines in this example. The "Hash", "Optimal" rows represent the result load distribution of each scheduling, respectively

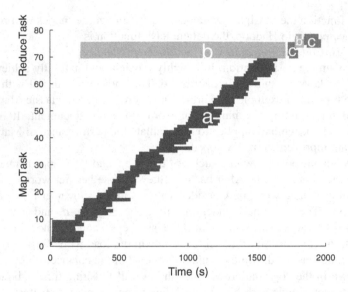

Fig. 4.3 We ran WordCount application on Amazon EC2 with 4 instances and we set 70 map tasks and 7 reduce tasks. This figure describes the timing flow of each *Map* task and *Reduce* task. Region *a* represents the actual map function executing time. Region *b* represents the shuffling time and region *c* represents the actual reduce function executing time. The regions between both *b* and *c* represent the sorting time

reduce phase further consists of three subphases: *shuffle*, in which the task pulls the map outputs; *sort*, in which the map outputs are sorted by keys; and *reduce function execution*, in which a user-defined function takes the map outputs with each key and, after all of the mappers have finished working, starts to run and generates the final outputs.

In Fig. 4.3, we show a detailed measurement of the processing times of all of the phases for a WordCount application running on Amazon EC2. The *map* phase starts at time zero. The *reduce* phase, and the *shuffle* subphase of the *reduce* then starts at about 200s. We can see that the shuffle finishing time is much longer than the reduce function executing time, because the reduce workers need to wait for the map workers to generate intermediate pairs while using remote procedure calls to read the buffered data from the local disks of the map workers. Also note that the maximum map finishing time is quite close to that of shuffle subphase. Therefore, if we wait until all of the keys are generated, in other words, if we start the *shuffle* subphase after all of the map workers have finished, the overall job finishing time will double. This is unfortunately what the state-of-the-art offline algorithms do for load balancing. It is what have motivated us to design an online solution to start the *shuffle* subphase as soon as possible, while making the maximum load of the reduce workers as low as possible.

4.2.2 Problem Formulation

We consider a general scenario where, during the map phase, the mapper nodes generate many intermediate values (data) with associated keys. Each of these, which we denote as (k_i, l_i), forms a (key, location) pair, where location l_i refers to where the (key, value) pair is stored, and the worker nodes report the pair to the master node.[1] In the rest of this chapter, we will be mainly interested in the issue of processing the *key* attribute of such a pair; for the sake of simplicity, we will usually refrain from discussing the location attribute.

The master node then assigns these pairs to different machines based on the key values. Each such pair must be assigned to one machine for processing, with the additional restriction that pairs with the same key must be assigned to the same machine. The number of pairs assigned to a machine makes up the *load* of the machine. Here, we assume that each machine will have a finishing time that is directly proportional to its load, and that the finishing time of the machine with the highest load (which we call the *makespan*) will be the overall finishing time.

The objective then is to minimize the overall finishing time by minimizing the maximum load of all of the machines.[2]

Formally, for the master node, the input is a stream $S = (b_1, b_2, \cdots, b_N)$ of length N, where each b_i denotes a (key, location) pair. Let N' denote the number of different keys; we denote $C = \{c_1, c_2, \cdots, c_{N'}\}$ as the universal set of the different keys, with $b_i \in C$ for every $i \in N$. We assume that there are m identical machines numbered $1, \ldots, m$. We denote the load of machine i by M_i, i.e., the number of pairs assigned to machine i. Initially, all loads are 0. Our goal is to assign each b_i in S to a machine, so as to obtain

$$\min_{i \in \{1, 2, \cdots, m\}} \text{Max } M_i$$

such that any two pairs (k_1, l_1) and (k_2, l_2) will be assigned to the same machine if $k_1 = k_2$.

4.2.3 Input Models

Basically, we see that this is a streaming problem, where we need to make online decisions to balance the loads. We consider two input models, leading to two computational models. In our first model, we allow arbitrary (possibly adversarial)

---

[1] Note that the value is *not* being reported, and thus, the information received by the master node for each item will only require a small amount of space.

[2] Here we make an implicit assumption that each pair represents a workload of unit size, but our algorithm can easily work also for variable integer workload weights.

input, and stipulate that the master node will assign the pairs in a purely online fashion. In our second model, we assume that the input comes from a probability distribution, and, in order to exploit this fact, the master node is allowed to store and process *samples* of its input before starting to assign the samples and the rest of the pairs in an online fashion. In the next two sections, we develop algorithms that follow each of these two models respectively.

4.3 A 2-Competitive Fully Online Algorithm

In order to minimize the overall finishing time, it makes sense to start the *shuffle* subphase as soon as possible, with as much of an overlap with the *Map* phase as possible. In this section, we give an *online* algorithm, List-based Online Scheduling, for assigning the keys to the machines. Our algorithm decides, upon receiving a (key, location) pair, which machine to assign that item to without any knowledge of what other items may be received in the future. We assume that the stream of items can be arbitrary, that is, after our algorithm makes a particular assignment, it can possibly receive the "worst" stream of items for that assignment. Our algorithm will be analyzed for the worst-case scenario: we will compare its effectiveness to that of the best offline algorithm, i.e., one that makes its decisions with knowledge of the entire input.

For assigning items to machines based on their keys, we adopt a Greedy-Balance load balancing approach [10] of assigning unassigned keys to the machine with the smallest load once they come in.

Algorithm 5 List-based Online Scheduling

1: Read pair (k_i, l_i) from S
2: **if** k_i has been assigned to machine j **then**
3:     Assign the pair to the machine j
4: **else**
5:     Assign the pair to the machine with the least load

We now show that this algorithm yields an overall finishing time that is at most twice that of the best offline algorithm.

Theorem 4.1. *List-based Online Scheduling has a competitive ratio of 2.*

Proof. Let OPT denote the offline optimum makespan, which is the maximum finishing time. Assume that machine j is the machine with the longest finishing time in the optimal offline solution and that T' is the number of pairs read just before the last new key, say c_j, is assigned to machine j. Obviously, T' must be less than N, the total length of the input. Then, at the time that c_j is assigned to machine j, j must have had the smallest load, say L_j. Thus we have:

$$L_j \leq \frac{T'}{m} < \frac{N}{m} \leq OPT$$

Let $|c_j|$ denote the number of pairs with key c_j in S. Then, the finishing time of machine j, denoted by T, which is also the makespan, is

$$T = L_j + |c_j| \leq OPT + OPT = 2OPT$$

Thus, with our List-based Online algorithm, the makespan can achieve a 2-competitive ratio to the offline optimal makespan.   □

4.4   A Sampling-Based Semi-online Algorithm

Our previous algorithm made no assumptions about our advance knowledge of the key frequencies. Clearly, if we had some *a priori* knowledge about these frequencies, we could make the key assignments more efficiently.

In this section, we assume that the pairs are such that their keys are drawn independently from an unknown distribution. In order to exploit this, we compromise on the online nature of our algorithm and start by collecting a small number of input pairs into a *sample* before making any assignments. We then use this sample to estimate the frequencies of the K most frequent keys in this distribution, and use this information later to process our stream in an online fashion. In order to observe the advantages of such a scheme, consider another toy example shown in Fig. 4.4. If we can wait for a short period before making any assignments, for instance, collect the first nine keys in the example and assign the frequent keys to the machine with the least load in order of frequency, the maximum load is reduced to 12 from 15.

Our algorithm classifies keys into two distinct groups: the K most frequent, called *heavy* keys, and the remaining, less frequent, keys. The intuition is that the heavy keys contribute much more strongly to the finishing time than the other keys, and thus need to be handled more carefully. As a result, our algorithm performs assignments of the keys to the machines differently for the two groups.

We first consider how to identify the heavy keys. Clearly, if one could collect all of the stream S, the problem can be solved exactly and easily. However, this would use too much space and delay the assignment process. Instead, we would like to trade off the size of our samples (and wait before we start making the assignments) with the accuracy of our estimate of the key frequencies. We explore the parameters of this trade-off in the rest of this section.

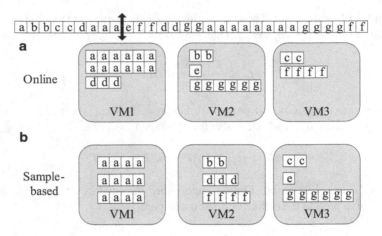

Fig. 4.4 An illustrative example showing benefits of sampling. The setting is the same as in Fig. 4.2. The "Online" and "Sample-based" rows represent the respective result load of each machine in the online and sample-based schedules

Our first goal is to show that we can identify the most frequent (heavy) K keys reliably. We first analyze the sample size necessary for this task using techniques from probability and sampling theory. We then move on to an algorithm for assigning the heavy keys as well as the remaining, less frequent, ones.

4.4.1   Sample Size

In this section we analyze what our sample size needs to be in order to obtain a reliable estimate of the key frequencies. Estimating probabilities from a given sample is well understood in probability and statistics; our proof below follows standard lines [5].

Let S' denote our sample of size n and n' denote the number of distinct keys in S'. For simplicity, we will ignore the fact that S' consists of (key, location) pairs, instead considering it as a stream (or set) of keys.

Let p_i denote the proportion of key c_i in the stream S, and let X_i denote the number of occurrences of c_i in the sample S'. (It is possible to treat p_i as a probability as well, without any changes to our algorithm.) Then X_i can be regarded as a binomial random variable with $E(X_i) = np_i$ and $\sigma_{X_i} = \sqrt{p_i(1 - p_i)n}$. Provided that n is large (i.e., $X_i \geq 5$ and $n - X_i \geq 5$), the Central Limit Theorem (CLT) implies that X_i has an approximately normal distribution regardless of the nature of the item distribution.

In order to select the heavy keys, we need an estimate of key probabilities. To this end, we estimate p_i as $\hat{p}_i = X_i/n$, which is the sample fraction of key i in S'. Since $\hat{p}_i$ is just X_i multiplied by the constant $1/n$, $\hat{p}_i$ also has an approximately normal distribution. Thus, $E(\hat{p}_i) = p_i$ and $\sigma_{\hat{p}_i} = \sqrt{p_i(1 - p_i)/n}$, and we have the following theorem bounding the size of the sample that we need in order to have a good estimate of the key frequencies.

Theorem 4.2. *Given a sample S' of the size of $n = (z_{\alpha/2}/\epsilon)^2$, consider any key c_i with proportion p_i, satisfying $X_i \geq 5$ and $n - X_i \geq 5$, and let $\hat{p}_i = X_i/n$. Then, $|p_i - \hat{p}_i| \leq \epsilon \sqrt{(\hat{p}_i)(1 - \hat{p}_i)}$ with a probability of $1 - \alpha$.*

Note that $z_{\alpha/2}$ is a parameter of the normal distribution whose numeric value depends on α and can be obtained from the normal distribution table.

4.4.2 Heavy Keys

We first state our notion of the more frequent (i.e., heavy) keys. The following guarantees that we will explore all keys with a length of at least $OPT/2$.

Definition 4.1 (Heavy Key). A key i is said to be *heavy* if $\hat{p}_i \geq 1/2m + \epsilon$.

Note, then, that a key whose length (i.e., the number of times that it occurs in S) is greater than $N/2m$ is very likely to be heavy. Then, it is easy to see that there could be up to $2m$ heavy keys.

It is worth noting that one might need to see $O(m)$ samples to sample a particular heavy key. Thus we will need to increase our sample size by an $O(m \log m)$ factor to ensure that we sample the heavy keys and estimate the lengths of each of the heavy keys reliably, resulting in a sample size of $n = O((z_{\alpha/2}/\epsilon)^2 m \log m)$.

4.4.3 A Sample-Based Algorithm

We are now ready to present an algorithm for assigning the heavy keys, similar to the sorted-balance algorithm [10] for load balancing. Our sample-based algorithm first collects samples. It then sorts the keys in the sample in a non-increasing order of observed key frequencies, and selects the K most frequent keys. Then, going through this list, it assigns each type of key to the machine with the least current load. For assigning all other keys, we develop a Sample-based Algorithm as follows.

Algorithm 6 Sample-based Algorithm

---

    wait until $n = O((z_{\alpha/2}/\epsilon)^2 m \log m)$ pairs are collected to form a sample
    sort the K most frequent keys in the sample in non-increasing order, say $\hat{p}_1 \geq \hat{p}_2 \geq \cdots \geq \hat{p}_K$
    going over the sorted list, assign each key i to the machine with the smallest load
    while a new pair is received with key i **do**
        if the i was previously assigned to machine j **then**
            assign i to machine j
        else
            assign i to the machine with the smallest load

The following lemma bounds the size of the last key assigned to the machine that ends up with the longest finishing time.

Lemma 4.1. *If the makespan obtained by the Sample-based algorithm is larger than $OPT + \epsilon N$, with a probability of at least $1 - 2\alpha$, the last key added to the machine has a frequency of at most $(OPT/2N + \epsilon)$.*

Proof. Let OPT be the optimal makespan of the given instance. Divide the keys into two groups: $C_L = \{j \in C : \hat{p}_j N > OPT/2 + \epsilon N\}$ and $C_S = C - C_L$, called large and small keys respectively. With probability $1 - \alpha$, we have $p_j N > \hat{p}_j N - \epsilon N > OPT/2$ for all keys j. Note that there can be at most m large keys, otherwise one could not obtain a finishing time of OPT with two such keys scheduled on the same machine. Since the length of a large key is greater than $OPT/2$, this contradicts the view that OPT is the optimal makespan. It is also obvious that we cannot have any keys with a length of greater than OPT, i.e., no j exists such that $p_j N > OPT$. Thus, if the makespan obtained by the algorithm is greater than $OPT + \epsilon N$, with probability $1 - \alpha$, the last new key that is assigned to the makespan machine must be a small key. Using the union bound, with probability $1 - 2\alpha$, the last type of key assigned to the machine with the longest processing time must have a frequency of at most $OPT/2N + \epsilon$. $\square$

Theorem 4.3. *With a probability of at least $1-2\alpha$, $0 < \alpha < 1/2$, the Sample-based algorithm will obtain an overall finishing time of at most $\frac{3}{2}OPT + N\epsilon$.*

Proof. Assume that machine j has the longest finishing time when the Sample-based algorithm is used for the key assignments. Consider the last key k assigned to j. Before this assignment, the load of j was L_j, and it must be the least load at that point in time among all of the machines. Thus,

$$L_j \leq \frac{N}{m} \leq OPT.$$

Then, after adding the last key k, its finishing time becomes at most

$$L_j + N p_k \leq OPT + OPT/2 + \epsilon N \leq \frac{3}{2}OPT + \epsilon N.$$

Note that $L_j \leq OPT$ is deterministically true. Therefore, the probability of the above can be shown to be at least $1 - 2\alpha$, $0 < \alpha < 1/2$. Thus, with at least $1 - 2\alpha$, $0 < \alpha < 1/2$, our Sample-based algorithm can achieve $\frac{3}{2}OPT + \epsilon N$. $\square$

4.5 Performance Evaluation

4.5.1 Simulation Setup

We evaluate our algorithms on both a real data trace and synthetic data. The real trace is a public data set [1], which contains the Wikipedia page-to-page link for each term. This trace has a data size of 1 Gigabytes. We generate the synthetic data according to a Zipf distribution with a varying parameter s, by which we can control the skew of the data distribution.

In our performance evaluation, we not only simulate the data assignment process, but also the procedures by which the reduce workers pull data from a specific place.

We evaluate both of our two algorithms: the List-based online scheduling algorithm (Online) and the Sample-based algorithm (Sample-based). Recall that the former is faster, while the latter has better accuracy. We compare our algorithms to the current MapReduce algorithm with the default hash function (Default). To set a benchmark for our Online algorithm, we also compare it to the offline version (Offline), which sorts the keys by their frequencies and then assigns them to the machine with the least load so far. The primary evaluation criteria for these algorithms are the maximum load and the shuffle finishing time.

The default values in our evaluation are $z_{\frac{a}{2}} = 1.96$, $\epsilon = 0.005$; the number of records is 1,000,000; and the number of identical machines is 20. The parameter s is set to 1 by default, and we also vary it to examine its impact. Note that we scale the y axis to make the figure visually clean.

4.5.2 Results on Synthetic Data

Figure 4.5 shows the maximum load as a function of the number of data records on the synthetic data. Our data record is the key, as mentioned earlier. We increase the number of data records from 0.5×10^6 to 2.3×10^6. We compare all four algorithms. We can see that the performance of the Default algorithm is much worse than that of the other three. When the number of data records is 2.1×10^6, the maximum load of the Default algorithm is 3.78×10^5 and our Online algorithm has a maximum load of only 2.3×10^5, an improvement of 39.15 %. In addition, we see when the number of data records increases, the maximum loads of all of the algorithms will also increase. This is not surprising, as we need to process more data. However, the loads in our algorithms increase at a much slower pace as compared to the Baseline algorithm. Further, we can see that the performance of our Online algorithm is almost identical to that of the Offline algorithm. This indicates our algorithm not only bounds the worse-case scenario theoretically, but also in practice performs much better than the theoretical bound.

Figure 4.6 shows the maximum load as a function of the reducer number on the synthetic data of all of the four algorithms. We increase the reducer number from

Fig. 4.5 Maximum load as a
function of data record
number (synthetic data)

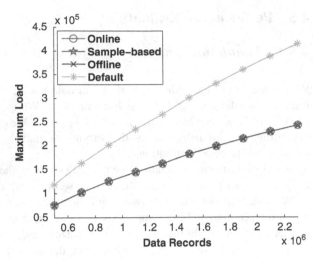

Fig. 4.6 Maximum load as a
function of reducer number
(synthetic data)

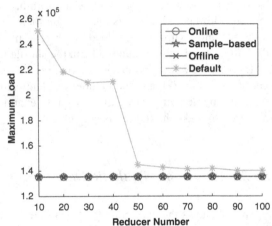

10 to 100. We can see that the Default algorithm performs much worse than the
other three. In particular, when the reducer number is 20, the maximum load of the
Default algorithm is 2.2×10^5, while the maximum load of our Online algorithm
is only 1.3×10^5, a reduction of 40.90 %. It is natural for the maximum load to
decrease as reducer number increases, as the Default algorithm shows. However,
it is interesting that the other three algorithms do not change much as the reducer
number increases. We have checked the data distribution and found that there is
one key that is extremely frequent. Our algorithms have indeed identified this key
so that the performance of our Online algorithm is almost identical to that of the
Offline algorithm.

Figure 4.7 compares the maximum load of all the four algorithms as a function
of the skew on the synthetic data. We set the skew by adjusting parameter s from
0.5 to 1.5 in the Zipf distribution function. The larger s is, the more skew the data

Fig. 4.7 Maximum load as a function of data skew (synthetic data)

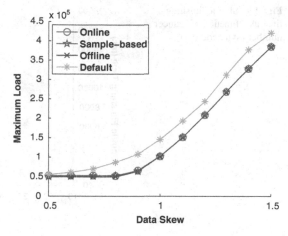

Fig. 4.8 Shuffle finishing time as a function of data record number (synthetic data)

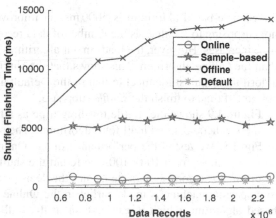

has. We can see that when the data are even, the parameter s ranges from 0.5 to 0.7; and the maximum loads of all of the four algorithms are almost the same. When the data distribution becomes increasingly skewed, it is easy to recognize that the Default algorithm behaves much worse than all of the other three. Not surprisingly, when the data are highly skewed, the balancing strategies are always better than that of the Default algorithm. This is because the maximum load is the frequency of the most frequent key. For the Online algorithm and the Sample-based algorithm, it is easy to identify the most frequent key without having to see all of the keys.

Figure 4.8 shows the shuffle finishing time as a function of the number of data records on the synthetic data of all of the four algorithms. We still vary the number of records from 0.5×10^6 to 2.3×10^6. We can see that the Offline algorithm behaves much worse than all of the other three algorithms. In particular, when the number of data records is 2.1×10^6, the shuffle finishing time of the Offline algorithm is 14,000 ms and our Online algorithm has a shuffle finishing time of 1000 ms, an improvement of 14 times. In comparison, the shuffle finishing time of

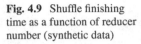

Fig. 4.9 Shuffle finishing time as a function of reducer number (synthetic data)

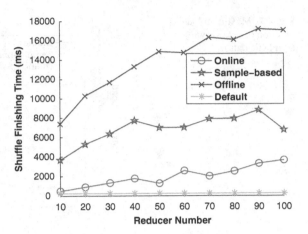

the Sample-based algorithm is 5000 ms, an improvement of almost 3 times. It is not surprising to see that, as the number of data records grows, the shuffle finishing time increases. However, the loads in our algorithms increase at a much slower pace than the Offline algorithm. This shows that the shuffle finishing time of our Online algorithm is almost identical to that of the Default algorithm, which takes the least amount of time to finish the *shuffle* subphase.

Figure 4.9 shows the shuffle finishing time as a function of the reducer number on the synthetic data with all four algorithms. The result is similar to that depicted in Fig. 4.8. We tested the performance of the Online algorithm by increasing the reducer number from 10 to 100. We found the shuffle finishing time of the Online algorithm to be good as expected, since the decision to assign a newly incoming key to a specific machine is made earlier in the Online algorithm than in the Sample-based algorithm, which is earlier than in the Offline algorithm. We should note that as the reducer number increases, the shuffle finishing time also increases for all of the algorithms expect for the Default algorithm. This is because we need to check whether the reducer machine contains the incoming keys or gets the least load machine in these algorithms.

Figure 4.10 shows the shuffle finishing time of all four algorithms as a function of data skew on synthetic data. We still set the skew parameter from 0.5 to 1.5. Note that the data sets are generated independently for the different skew parameters, with the result that the trend of the shuffle finishing time in each algorithm is not as monotonic as expected. This result is similar to that depicted in Fig. 4.9.

4.5.3   Results on Real Data

Figure 4.11 shows the maximum load as a function of the number of (key, value) pairs in the real trace dataset. We tested the performance by setting the number of

Fig. 4.10 Shuffle finishing time as a function of data skew (synthetic data)

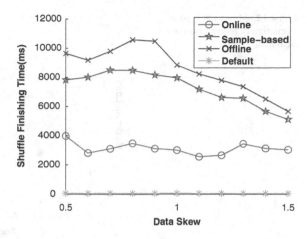

Fig. 4.11 Maximum load as a function of data record number (real data trace)

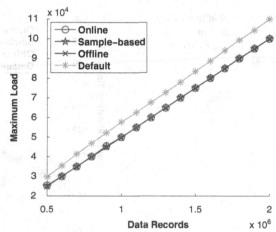

(key, value) pairs from 0.5×10^6 to 2.0×10^6. Our algorithm always performs better than the Baseline algorithm. In particular, when the number of (key, value) pairs is 900,000, the maximum load of the Baseline algorithm is 16,993 unit sizes, while the maximum load of our algorithm is 9002 unit size, an improvement of 47.03 %. A similar result is found in Fig. 4.5 on the synthetic data.

Figure 4.12 shows the maximum load as a function of the number of reducers on both algorithms in the real trace dataset. We evaluated the performances by increasing the reducer number from 100 to 190. We can see the performance of Baseline algorithm is much worse than that of our algorithm. It is natural for the maximum loads of all of the algorithms to decrease when the reducer number increases. Further, we can see that the performance of our Sample-based algorithm is almost identical to that of the Offline algorithm.

Figure 4.13 compares the shuffle finishing time of all four algorithms as a function of the number of data records in the real trace dataset, which is similar to

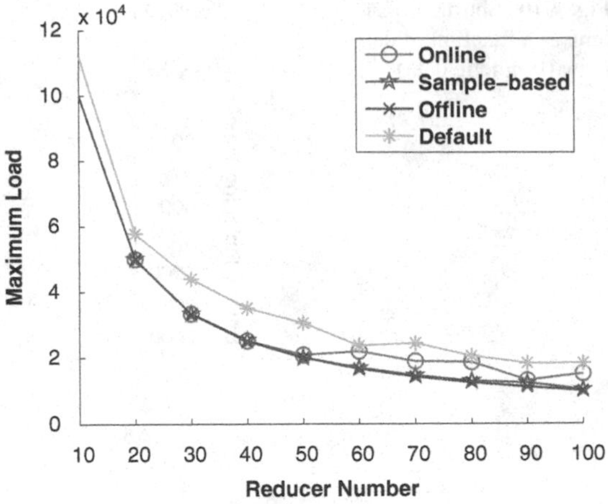

Fig. 4.12 Maximum load as
a function of reducer number
(real data trace)

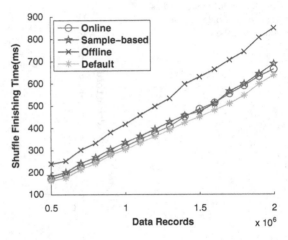

Fig. 4.13 Shuffle finishing
time as a function of data
record number (real data
trace)

the result as shown in Fig. 4.9. Our Online algorithm, the Sample-based algorithm, and the Default algorithm achieved better results than that of the Offline algorithm. In addition, we found that when the number of data records increases, the maximum loads of all of the algorithms increase. This is because more time is needed to process more data. However, the loads in our algorithms increased in a much slower pace as compared to the Offline algorithm. This is because when the data records become larger, a longer processing time is required in the map phase, making the Offline algorithm wait much longer.

Figure 4.14 shows the shuffle finishing time as a function of the reducer number of all four algorithms. We also found that the shuffle finishing time grows as the number of reduce workers increases. This is illustrative, since it will cost much more time to check whether or not the incoming key was assigned and to find machine with the least load. Interestingly, the shuffle finishing time of the Offline algorithm increases much faster than the other three. This gap probably appears

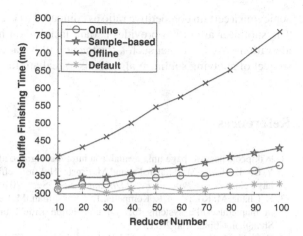

Fig. 4.14 Shuffle finishing time as a function of reducer number (real data trace)

because the overall waiting time and the increased cost brought by the increasing number of reducers is very large.

In summary, the performance of our Online and Sample-based algorithm perform close to that of the Offline algorithm in terms of finishing time. The two algorithms consistently perform better than the MapReduce Default algorithm from a maximum load point of view. Our algorithms also show a comparable shuffle finishing time to that of the Default algorithm, and are better than the Offline algorithm in that regard.

4.6 Summary and Discussions

In this chapter, we presented one problem related to big data processing. More specifically, we faced a data skew problem in the shuffling phase of MapReduce. We observed that the data skew problem can be translated into a heavy hitter problem. This is where the sublinear algorithm is applied. Such a linkage can be common when applying sublinear algorithms to real-world applications.

The original sorted-balance algorithm can achieve $\frac{4}{3}OPT$ [10] in the maximum finishing time. Our semi-online algorithm achieved $\frac{3}{2}OPT$ plus some additive error for the K most frequent keys. A further shrinking of this gap could be achieved through advanced algorithm design, or the error could be further reduced. The keys could also be finely classified into different groups according to certain weights, so as to refine the results. In addition, if the distribution of data follows a known distribution, e.g., the Zipf distribution, the parameters can be better estimated, making the process of the identifying the K initial keys much easier and more accurate. It should also be possible to make the additive error smaller as well.

Note that the sublinear algorithm described in this section only serves to assist the whole algorithm. We have basically tried to solve the problem formulated in Sect. 4.2. To achieve this goal, we developed online algorithms with the target of achieving a certain competitive ratio as compared to the respective offline algorithm. The sublinear algorithm provides certain insights on how to develop better online algorithms. We hope that such a treatment will broaden the view of readers on the subject of applying sublinear algorithms to real applications.

References

1. Wikipedia page-to-page link, available at http://haselgrove.id.au/wikipedia.htm.
2. Y. Bu, B. Howe, M. Balazinska, and M. Ernst, "HaLoop: efficient iterative data processing on large clusters", in *Proc. of the VLDB Endowment*, Sept. 2010.
3. H. Chang, M. Kodialam, R. Kompella, T. V. Lakshman, M. Lee, and S. Mukherjee, "Scheduling in mapreduce-like systems for fast completion time", in *Proc. of IEEE INFOCOM'11*, Shanghai, China, Apr. 2011.
4. F. Chen, M. Kodialam, and T. V. Lakshman, in *Proc. IEEE INFOCOM'12*, "Joint scheduling of processing and Shuffle phases in MapReduce systems", Orlando, Florida, USA, Mar. 2012.
5. J. Devore, *Probability & Statistics for Engineering and the Sciences*, CengageBrain.com, 2012.
6. J. Ekanayake, H. Li, B. Zhang, T. Gunarathne, S. Bae, J. Qiu, and G. Fox, "Twister: a runtime for iterative MapReduce", in *Proc. ACM HPDC'10*, Chicago, Illinois, USA, June, 2010.
7. M. Englert, D. Ozmen, and M. Westermann, "The Power of Reordering for Online Minimum Makespan Scheduling", in *Proc. IEEE FOCS'08*, Philadelphia, Pennsylvania, USA, Oct. 2008.
8. B. Gufler, N. Augsten, A. Reiser, and A. Kemper, "Handling Data Skew In MapReduce", in *The First International Conference on Cloud Computing and Services Science*, 2011.
9. B. Gufler, N. Augsten, A. Reiser, and A. Kemper, "Load Balancing in MapReduce Based on Scalable Cardinality Estimates", in *Proc. IEEE ICDE'12*, Washington, DC, USA, Apr. 2012.
10. J. Kleinberg and E. Tardos, *Algorithm Design*, Pearson Education India, 2006.
11. Y. Kwon, M. Balazinska, B. Howe, and J. Rolia, "A study of skew in mapreduce applications", in *The 5th Open Cirrus Summit*, 2011.
12. Y. Kwon, M. Balazinska, B. Howe, and J. Rolia, "SkewTune: Mitigating Skew in MapReduce Applications", in *Proc. ACM SIGMOD'12*, Scottsdale, Arizona, USA, May. 2012.
13. W. Lang and J. Patel, "Energy management for MapReduce clusters", in *Proc. of the VLDB Endowment*, Sept. 2010.
14. J. Leverich and C. Kozyrakis, "On the energy (in) efficiency of Hadoop clusters", in *ACM SIGOPS Operating Systems Review*, Jan. 2010.
15. B. Li, E. Mazur, Y. Diao, A. McGregor, and P. Shenoy, "A platform for scalable one-pass analytics using MapReduce", in *Proc. ACM SIGMOD'11*, Athens, Greece, June, 2011.
16. J. Lin, "The Curse of Zipf and Limits to Parallelization: A Look at the Stragglers Problem in MapReduce", in *The 7th Workshop on Large-Scale Distributed Systems for Information Retrieval*, July. 2009.
17. G. Malewicz, M. Austern, A. Bik, J. Dehnert, I. Horn, N. Leiser, and G. Czajkowski, "Pregel: a system for large-scale graph processing", in *Proc. ACM SIGMOD'10*, Indianapolis, Indiana, USA, June, 2010.
18. K. Morton, M. Balazinska, and D. Grossman, "ParaTimer: a progress indicator for MapReduce DAGs", in *Proc. ACM SIGMOD'10*, Indianapolis, Indiana, USA, June. 2010.
19. S. Ramakrishnan, G. Swart, and A. Urmanov, "Balancing reducer skew in MapReduce workloads using progressive sampling", in *Proc. ACM SoCC'12*, San Jose, California, USA, 2012.
20. M. Schatz, "CloudBurst: highly sensitive read mapping with MapReduce", in *Bioinformatics*, vol. 25, no. 11, pp. 1363–1369, 2009.
21. J. Stamos and H. Young, "A symmetric fragment and replicate algorithm for distributed joins", in *IEEE Transactions on Parallel and Distributed Systems*, 1993.

22. J. Tan, X. Meng, and L. Zhang, "Coupling task progress for MapReduce resource-aware scheduling", in *Proc. IEEE INFOCOM'13*, Turin, Italy, Apr. 2013.
23. W. Yan and P. Larson, "Eager Aggregation and Lazy Aggregation", in *Proc. VLDB'95*, Zurich, Switzerland, Sept. 1995.
24. H. Yang, et. al., "Cloud 9: A MapReduce library for Hadoop, available at http://lintool.github.io/Cloud9/
25. H. Yang, A. Dasdan, R. Hsiao, and D. Parker, "Map-reduce-merge: simplified relational data processing on large clusters", in *Proc. ACM SIGMOD'07*, Beijing, China, June, 2007.
26. M. Zaharia, A. Konwinski, A. Joseph, R. Katz, and I. Stoica, "Improving MapReduce Performance in Heterogeneous Environments", in *Proc. USENIX OSDI'08*, Dec. 2008.

Chapter 5
Application on a Smart Grid

5.1 Introduction

A smart grid introduces novel concepts on energy efficiency and conservation with the benefit of smart meters. Smart grids have gained worldwide recognition in the past few years. The interpretation of "smart" reflects a noticeable change in the way energy is generated, distributed, delivered, and consumed. Integrating advanced electrical and communications infrastructures incorporated with modern process automation techniques, a smart grid system is constructed in such a way that two-way communication between electricity suppliers and residential electricity consumers is efficiently implemented. An illustration of a smart grid system is given in Fig. 5.1. With these powerful characteristics brought about by modern electrical technologies, a smart grid significantly alters the way utility companies, governments, customers, and business participants view electricity transmission and its associated services.

One of the core identities in a smart grid is the smart meter. Different from the traditional electromechanical electric meter, a smart meter is defined by various functions, such as: two-way communications between the utility and the meter; the constant recording of power consumption at an interval such as 15 min; the integration of a home area network (HAN) interface; the detection of power quality; and the reliable and secure data exchange. The design of a smart meter helps to overcome the drawbacks of traditional electromechanical meters, such as inefficiency, inaccuracy, inclined tampering, the lack of a means for carrying out remote monitoring or control, and the absence of a way for the consumer to view his or her energy usage. The popular employment of smart meters mainly relies on the use of advanced metering infrastructure (AMI) systems. AMI systems consist of communication systems, consumer energy displays and controllers, software applications, meter data management systems and supplier business systems. The network allows information such as time-of-use pricing information to be collected and distributed to customers, utility companies, and service providers. The major

© The Author(s) 2015
D. Wang, Z. Han, *Sublinear Algorithms for Big Data Applications*,
SpringerBriefs in Computer Science, DOI 10.1007/978-3-319-20448-2_5

User   Utility company   Power Generator

Fig. 5.1 Illustration of smart grid system

benefit of AMI is the reduction in costs compared to a manual system. Some other outstanding advantages of AMI include: valuable outage detection, tamper and theft detection, demand management, and comprehensive customer information.

With the incorporation of electrical devices and data recording components, smart grid systems all over the world are consuming a tremendous amount of data from time to time. Due to the enormous amount of electricity consumed, analyzing the big data produced by smart meters is a crucial challenge for electricity companies and researchers. In particular, the detailed use of electricity in households is recorded by the smart meter in such a way that it is possible to analyze the power usage behaviors for the purpose of devising better pricing schemes. Information on power usage is valuable in that it provides guidance to utility companies or governments on how to regulate the consumption and allocation of power for energy conservation and sustainability. One of the fundamental issues regarding user behaviors is to understand typical patterns of behavior. The problem involves determining features and classifying patterns based on usage data. In practice, user load profiles are usually selected as the feature to characterize users. A load profile is a graph of the variation in the electrical load versus time. Various factors influence the shape of the load profile, including customer type (for instance, residential, commercial, or industrial), temperature, and holidays. Viewed from different time units (day/month/season/year), the load profile show patterns of repetition in different time segments, indicating the dimensionality of the power usage data as well as the distribution of the load profile. As a consequence of selecting such features for investigation, the data under consideration can be substantially reduced. It is in the area of addressing the challenge of processing the data in a smart grid that the sublinear algorithms can be applied. A promising approach is to use sublinear algorithms, to classify users based on only some portion of sub-sampled data points, instead of processing the entire usage data set, which would be prohibited in computation.

The emergence of smart grids has been considered a potential boost for better energy management, as a result of two-way communications. The utility companies can group users by their behaviors based on the collected usage information, and thereby, provide sophisticated user-oriented services. With the foundation laid by analyzing patterns of smart meter data, differentiated services are now feasible and attracting increasing attention. The underlying philosophy of differentiated services is quite straightforward: for different types of users, electricity prices should differ. Compared with traditional fixed/static pricing schemes, differentiated services are more attractive. From the point of view of users, more suitable charge plan can be found within user-oriented pricing schemes. From the point of view of utility companies, user behaviors can be regulated by adjusting pricing and better demand management can be achieved. Dynamic pricing is one popular example of differentiated services. The implementation of dynamic electricity rates based on the smart meters is one way to influence consumers: for instance, by setting higher rates for peak hours and lower rates for off-peak hours, dynamic pricing can potentially reduce peak loads. One critical issue with respect to differentiated services is that certain kinds of statistical calculations and estimations of power usage data are required in the construction and determination of pricing strategies, which involves heavy computation. In differentiated services settings, not only is the amount of data on the power usage of users is massive, but the number of power consumers under examination is also formidable. Again, this problem illustrates a proper application of sublinear algorithms. Employing sublinear algorithm in the differentiation of services would greatly reduce the computation load through the sub-sampling of only a tiny portion of the total population. Moreover, the theoretical bounds guaranteed by sublinear algorithms will also yield satisfying estimation results for the differentiated model.

5.1.1 Background and Related Work

Recently, smart grids have been attracting a great deal of attention from researchers. A thorough summary on the development, problems, and applications of smart grids is given in [8]. Authors [1] have discussed modern technologies for delivering power. The privacy and security issues associated with smart grids are the focus of the work in [4]. Smart grid technologies are investigated in [11] with an emphasis on communication components and the related standards.

Viewed from the perspectives of data analysis and model construction, many studies are dedicated to the topics of classifying user behaviors and pricing strategies. Authors in [9] employ fuzzy c-means clustering to disaggregate and determine energy consumption patterns from smart meter data. In [19], researchers propose a day-ahead pricing scheme taking user reactions and dynamic adjustments of price into account. Work in [15] provides a sophisticated model to address the evolution of energy supply, user demand, and market prices under real-time pricing in a dynamic fashion. There are also works involving game theory in smart grid applications, such as [18], where the case of only one supplier and multiple users is studied. It is worth pointing out that in previous studies a common approach was to devise dynamic pricing arrangements from the perspective of time intervals, i.e., where different hours/seasons are treated differently in the pricing model. However, the different behaviors of users are rarely discussed in the previous studies. In this chapter, we will take a close look at the problem of how to design a pricing scheme that differentiates users.

Viewed from the numerical computation perspective, sublinear algorithms have been heavily studied in connection with a recently emerging topic, big data [7]. Faced with heavy volumes and a wide diversity of data, the focus of research on big data is on exploring efficient approaches to accomplish tasks with little computation cost. The sublinear algorithm proposed in [14] introduces a novel way to make approximates using only a small portion of the entire data to obtain the result with guaranteed error bounds. The benefit of efficient computation comes at the cost of sacrificing accuracy under acceptable constraints. In [5], the authors investigate the use of a sublinear algorithm to estimate the weight of an Euclidean Minimum Spanning tree. Various sublinear algorithms have been developed to address problems such as seeking quantiles of data [17] and checking the periodicity of a given data stream [6]. Authors in [2] propose sublinear algorithms that can check the closeness of two given distributions. The outstanding property of the proposed algorithms is model-free universality because there are no prior assumptions about the structure of given distributions. Although sublinear topics have been widely studied in big data settings, they are rarely discussed in relation to the smart grids. In this chapter, we will take a further step than [2] and modify the sublinear algorithms so that they can be applied to classification problems in smart grids.

5.1.2   Chapter Outline

In this chapter, we study the application of sublinear algorithms in smart grids. First, we investigate the feasibility of employing differentiated services in a smart grid by conducting a comprehensive data analysis of smart meter data collected in the Houston area. We then take a deeper look at the user usage data to classify the load profiles employing sublinear algorithms. Armed with the foundation of load profile classifications, we investigate differentiated services, implementing the sublinear algorithms to enhance our computation. In addition, we evaluate for the performance of differentiated services by carrying out both a theoretical analysis and experiments using simulated data sets.

Below, in Sect. 5.2, we first present our analysis of the smart meter data, where we discuss the missing data problem and carry out a data trace study to reveal the statistical properties of the collected smart meter data. Section 5.3 is devoted to our proposed sublinear algorithms for classification with proven performance bounds. We construct the differentiated services based on different types of users in Sect. 5.4. In Sect. 5.5, we draw conclusions and present a further summary of this application.

5.2 Smart Meter Data Analysis

In this section, we first address the problem of completing missing smart meter data and then proposed a model to characterize user electricity usage behaviors.

5.2.1 Incomplete Data Problem

When collected data contains corrupted parts with missing variables or messy codes for some users, this is referred to as the incomplete data problem. Incomplete data are a commonly encountered problem in the industry. In a smart grid there may be many reasons for the problem, for example, broken data aggregation devices or electronic noise echoes in the transmission components. It is usually not a good idea to ignore incomplete data points since the missing values may encode important user information that could be crucial for the subsequent analysis. To some extent, the missing values can be completed at a price. Regarding the categories of incomplete data, it has been reported [12] that there are three main ways by which data might be missed: missing completely at random (MCAR), which means that the probability of data going missing is independent of instances; missing at random (MAR), meaning that the probability of data going missing is dependent on the observed variables; not missing at random (NMAR), indicating the probability of data going missing depends on the variables or the missing values.

There are various approaches to imputing the missing data, including deletion methods, maximum likelihood inference, and multiple imputation (MI). A mixture of models are employed to complete the missing data using maximum likelihood estimation in [10]. Another example is the multiple imputation through chained equations (MICE) model based on linear regressions, which offers the advantages of simplicity and efficiency. Moreover, the MICE model takes the relationship between the variables into consideration, making it suitable for real-world applications. More specifically, the overall procedure of MICE can be summarized as follows:

1. Delete the observations if every variable is missing;
2. For the rest of the missing observations, start imputation with randomly fill-in values drawn from the observed values;
3. Move through each type of variable and perform single variable imputation using linear regression;
4. Replace the originally random draws with the imputed values from the regression model and repeat Step 3);
5. Repeat Step 1) to Step 4) a certain number of times and create the multiple imputation data set. Average over the data set to obtain the final result.

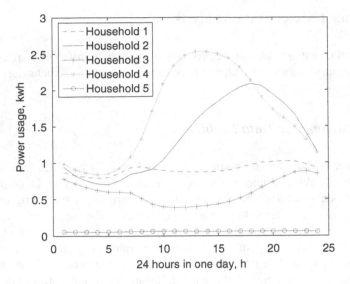

Fig. 5.2 Illustration of typical user load profiles

5.2.2 User Usage Behavior

A total number of about 2.2 million smart meters record power consumption in the Houston area with a time interval of 15 min. User load profiles are often characterized using data points of 24 dimensions, one dimension for each hour to represent power usage within a day. The load profile captures a user's power usage behavior. For the purpose of illustration, several load profiles of different types of households are presented in Fig. 5.2. As can be seen, household 1 represents the user group that has breakfast, leaves home, and then comes back home at dinnertime, while household 2 represents the user group that continues to use power when working in the house after breakfast. Household 3 represents users who behave similarly to those of household 2, but with a greater difference in power use between the afternoon and other hours. Household 4 represents users who mainly use energy from dawn to sunset. Household 5 represents users who consume little power compared with others.

In this study, users are classified according to their electricity usage distribution. A distribution is defined as a probability density function of a continuous/discrete random variable, which describes the likelihood that this random variable will take on a given value. It is true that there are many ways to characterize a user; for example, according to the total or average amount of electricity consumed in a month, the peak hour electricity usage on weekends/weekdays, and so on. We select electricity usage distribution as the feature for characterizing users because we believe that it provides a full spectrum of user electricity usage.

Formally, let **x** be a multi-dimensional random variable representing the daily load profile of a user. The usage distribution is defined as:

Definition 5.1. The electricity usage distribution $P\{\mathbf{x}\}$ is a distribution of the daily load profile **x**.

Based on an analysis of real smart meter data, we discover that many users have group properties in the sense that their distributions are similar, even though the exact usage of each user differs. This validates the choice of usage distribution as the criterion for classification. To abstract the usage distribution of each category, we choose to use a benchmark distribution, which is defined as:

Definition 5.2. A benchmark distribution is an electricity usage distribution $P\{\mathbf{x}\}$ with the expectation $\bar{\mathbf{x}} = E\{\mathbf{x}\}$, such that each $\bar{\mathbf{x}}_i$, $i = 1, \ldots, D(S)$ is a fixed value derived from real statistical data.

In Fig. 5.2, we see different average daily load profiles of various benchmark distributions. Differences in peak hours and peak usage are apparent among the plotted benchmark distributions. In our analysis of real data, there exist some cases where the average daily load profiles of some users are close, but there are noticeable differences in their usage distributions. For example, some users who consume no electricity at all during the weekends may still have similar average daily load profiles to those who constantly use power every day. We also determined that the seasonal effects have a great influence on user behaviors, resulting in similar average daily load profiles but different usage distributions. Moreover, we discovered that each dimension of the usage variable for an individual user conforms approximately to a Gaussian distribution. Users with similar average daily behaviors may end up with close Gaussian means but different variances in each dimension. All of these validate the choice of usage distribution to classify users. In order to classify users by their electricity usage distributions, we utilize the benchmark distributions with parameters predefined by the utility company.

5.3 Load Profile Classification

In this section, a well-studied sublinear algorithm is discussed in a case of testing whether two distributions are close. Based on this sublinear algorithm, we introduce a novel method for classification using distributions.

5.3.1 Sublinear Algorithm on Testing Two Distributions

There are many applications that involve distinguishing between two distributions: in computer science, for instance, we may need to differentiate two data streams by looking into their distributions on integers; in biology, it is often the case that we have to compare two sequences of genomes; in signal processing, various efforts have been dedicated to paralleling two time sequence signals. As the problem of big data is growing, the amount of data that we are coping with in those applications have become so huge that we need advanced, computational efficient approaches. One potential solution to this problem is the sublinear algorithm proposed in [2]. The essence of the proposed algorithm is to use a small portion of data to compute the results with a guaranteed error bound and a confidence parameter.

In detail, define two discrete distributions over n elements in the form of a probability vector as $\mathbf{p} = (p_1, \ldots, p_n)$ and $\mathbf{q} = (q_1, \ldots, q_n)$, where p_i is the probability of sampling the i-th element, and similarly for q_i. The objective is to test the closeness of these two distributions in the L_2-distance. The traditional way of doing this is to compute the exact value of the L_2-distance between the entire distribution. This is computationally inefficient in big data settings. Rather than employing direct calculation, the sublinear algorithm proposed in [2] takes a sub-sampling of the two distributions. The closeness of the two distributions is then determined based on some computed metrics. The algorithm repeats the whole procedure iteratively and outputs the final estimate. The two metrics used for measuring the similarity between two distributions are: (1) the collision probability, defined as the probability that a sample from each of $\mathbf{p}$ and $\mathbf{q}$ will yield the same element; this equals to $\mathbf{p} \cdot \mathbf{q}$; and (2) the self-collision of $\mathbf{p}$ and that of $\mathbf{q}$, defined similarly as $\mathbf{p} \cdot \mathbf{p}$ and $\mathbf{q} \cdot \mathbf{q}$, respectively. For convenience, the proposed sublinear algorithm is summarized in Algorithm 7.

Algorithm 7 Closeness testing based on L_2-distance

---

    for $i = 1, 2, \ldots, O(\log(1/\delta))$ **do**
        Let F_p = a set of m samples from $\mathbf{p}$.
        Let F_q = a set of m samples from $\mathbf{q}$.
        Let r_p be the number of pairwise self-collisions in F_p.
        Let r_q be the number of pairwise self-collisions in F_q.
        Let Q_p = a set of m samples from $\mathbf{p}$.
        Let Q_q = a set of m samples from $\mathbf{q}$.
        Let s_{pq} be the number of collisions between Q_p and Q_q.
        Denote $r = \frac{2m}{m-1}(r_p + r_q)$.
        Denote $t = 2s_{pq}$.
        if $r - t > m^2\epsilon^2/2$ **then**
            reject, i.e. consider the two distributions are different.
    Reject if the majority of the iterations reject, accept otherwise.

The intuition of the algorithm is as follows: r represents the sum of the self-collisions of the two individual distributions, and t represents the pair-wise collisions of the two distributions. If the difference between r and t is big, then the two distributions are not close.

In the algorithm, δ is also the parameter that determines the number of times that the algorithm has to be executed iteratively. A smaller δ imposes a larger iteration number. The parameter m represents the number of sub-sampled points from the distribution. To make the algorithm bounded, m has to be set at least $O(\epsilon^{-4})$. As proven in [2], the error and confidence parameters of Algorithm 7 are guaranteed by the following theorem.

Theorem 5.1. *Given ϵ, δ and distributions $\mathbf{p}$ and $\mathbf{q}$, Algorithm 7 on testing closeness passes with a probability of at least $1 - \delta$ if $\|\mathbf{p} - \mathbf{q}\| \leq \epsilon/2$, and with a probability of less than δ if $\|\mathbf{p} - \mathbf{q}\| > \epsilon$. The running time is $O(\epsilon^{-4} \log(1/\delta))$.*

Besides the computational advantage, Algorithm 7 makes no requirements for models for the distributions to be tested. This model-free character is what makes the proposed sublinear algorithm powerful in terms of generalization.

5.3.2 Sublinear Algorithm for Classifying Users

On the matter of classifying users by their usage distributions, a feasible approach is to classify users according to some benchmark distributions that represent typical user behaviors, as discussed in Sect. 5.2.2. Given a benchmark distribution and a test user, if the user's usage distribution is close to the benchmark, the user will be labeled as belonging to the category representing by the benchmark distribution. One weakness of Algorithm 7 is that when it is applied for classification, the confidence parameter is undetermined if $\mathbf{p}$ and $\mathbf{q}$ lie in the interval $[\epsilon/2, \epsilon]$ (see Theorem 5.1). In our L_2-distance-based classification, we need to set a threshold to classify users, i.e., if the distance between the user usage distribution and the benchmark distribution is below the threshold, the user will be labeled as belong to Group 1, indicating the same category represented by the benchmark; otherwise, the user will be labeled as belong to Group 0.

In this subsection, we develop a sublinear algorithm based on [2], but we give the complete confidence estimates. The essence of our proposed algorithm is to utilize Algorithm 7 twice, but with different parameters ϵ and 2ϵ, which helps to remove the interval of the undetermined confidence. Each time that the Algorithm 7 is called, the classified labels of the output are partially retained. Both partially retained results are then combined with some treatment to the overlapping labels, in order to obtain a final labeled result that is complete and consistent. We summarize the details of our proposed sublinear algorithm in Algorithm 8. It can be proven that with the parameter settings specified as in Algorithm 8, the accuracy of the final output can be guaranteed by the following lemma:

Lemma 5.1. *Given ϵ_2, δ_2 and distributions $\{\mathbf{p}_i\}$ and $\mathbf{q}$, the SubDist() of classifying users is based on the L_2-distance criteria: label user as 1 if $\|\mathbf{p}_i - \mathbf{q}\| \leq \epsilon_2$; label user as 2 if $\|\mathbf{p}_i - \mathbf{q}\| > \epsilon_2$. The classification accuracy is at least $1 - 2\delta_2$. In addition, $Pr[\text{labeled as } 1 | \text{true } 1] \geq (1 - 2\delta_2)$ and $Pr[\text{labeled as } 2 | \text{true } 2] \geq (1 - 2\delta_2)$.*

Algorithm 8 Modified sublinear algorithm based on L_2-distance testing

---

for each user's usage distribution $\mathbf{p}_i$ with a given fixed benchmark distribution $\mathbf{q}$ **do**

    *Step*1 : Employ Algorithm 7 with paramters $(\mathbf{p}_i, \mathbf{q}, m, \epsilon, \delta)$ and obtain the classification results as $\{LabelSet1\}$.

    *Step*2 : Employ Algorithm 7 with paramters $(\mathbf{p}_i, \mathbf{q}, m, 2\epsilon, \delta)$ and obtain the classification results as $\{LabelSet2\}$.

    *Step*3 : Keep the labeled 1 in $\{LabelSet1\}$ and reject all the labeled 0.

    *Step*4 : Keep the labeled 0 in $\{LabelSet2\}$ and reject all the labeled 1.

    *Step*5 : Combine the retained labels into $\{LabelSet3\}$; If the same user is both labeled as 1 in $\{LabelSet1\}$ and labeled as 0 in $\{LabelSet2\}$, his/her label is randomly determined as either 1 or 0 in $\{LabelSet3\}$.

    *Step*6 : Output $\{LabelSet3\}$ as the final classification results.

5.4 Differentiated Services

With advances in information and communications technologies integrated in a smart grid, we are able to record fine-grained information on user power usage for further analysis. Such information makes it possible for a utility company to provide various pricing plans for different types of users, with the help of a pattern analysis of smart meter data. Many utility companies are now devoting their efforts to developing differentiated services, not only because of the potentially greater profits to be derived from differentiated services but also for the sake of advanced demand-side management and better resource allocation.

Differentiated services are defined as services that vary in charge according to different types of objects. Generally speaking, in a smart grid field, the objects can be categorized as human factors and nonhuman factors. Regarding human factors, for instance, a utility company would probably charge personal and business users differently since they are likely to have different usage patterns. As for nonhuman factors, a utility company would probably adjust its price according to different hours of the day because users commonly consume more energy at noon and at night than at other times. Besides the factor of time, the factor of geography is also considered when a utility company settles down its pricing schemes for big cities and rural places.

One major objective for a utility company is to maximize its profit, defined as its revenue minus its costs. The revenue of a utility company depends on its pricing scheme. Many studies have treated differentiated services as a function of nonhuman factors. For example, the pricing scheme proposed in [16] is expressed in terms of piecewise linear functions whose pricing rates remain as different constant values in different time intervals within 24 h. However, differentiated services that take human factors into consideration are rarely discussed. In this section we focus on services in which users are differentiated on the basis of factors other than nonhuman ones. We describe our proposed differentiated services model in terms of a strategy that we have designed for the utility company: (1) classify different users based on benchmark distributions as discussed in Sect. 5.3; (2) set different pricing rates for classified user groups. The types of users and the pricing rates are two key factors in our proposed differentiated services model. These two factors can be obtained through optimization methods [13], engaging in gaming with other utility companies [3], or by addressing other external concerns of the utility company [11].

When estimating the profit for the utility company, the computation was seen to be over-burdened due to the big data arising from the huge number of users. Worse, such a computation would need to be executed repeatedly if optimization methods are employed to obtain pricing rates. Hence, we take a further step to estimate the expected profit instead of computing the exact value. The expected profit is estimated by replacing the individual usage pattern with the corresponding benchmark distribution. We compute the profit expected from a user group of certain type multiplied by the total number of users, the percentage of the given type of user, and the bill charged as a function of the given benchmark distribution. Thus, the total expected profit is the sum of the profit expected from each user group. We are particularly interested in percentage values because these can be utilized to quickly estimate income from bills without considering each user's information. In addition, percentage values can model the feedback from users. By comparing the percentage values of different years, the utility company can get an idea of how its past pricing schemes have affected user usage behaviors. The company might then revise its current pricing schemes in a dynamic fashion. In this way, power consumption can be regulated and more profits can be achieved. To alleviate the computational burden, the sublinear algorithm can be utilized again to estimate the percentage values and output the results with guarantees. Employing a sublinear algorithm to calculate the percentage values is straightforward: we iteratively take a sub-sampling of users from the total user pool and perform the computation using only the information on the sub-sampled users.

5.5 Performance Evaluation

In this section, we evaluate our proposed sublinear algorithm using numerical simulations. We also show the results of the impact of choosing different values of m as the sub-sampled number of data points.

The phantom data set is simulated based on an analysis of real data. For simplicity, we simulate binary types of users. The total number of users is set at $N = 100,000$. The percentage value of the type 1 user, α, varies from 0.1 to 0.8. Given two different benchmark distributions, the user usage distributions are generated from multivariate Gaussian distributions with different mean vectors and covariance matrices.

We define the estimation error as the absolute value of the difference between the estimated $\hat{\alpha}$ and the true α. By inputting the data set into our proposed sublinear algorithm with fixed parameters $\epsilon = 0.05$, $\delta = 0.05$, $m = 60$, we obtain the results shown in Fig. 5.3. As can be seen, our algorithm estimates the α values precisely within the error bounds throughout all of the simulated values. And the sub-samples that are used comprise only about 1/6 of all distribution points.

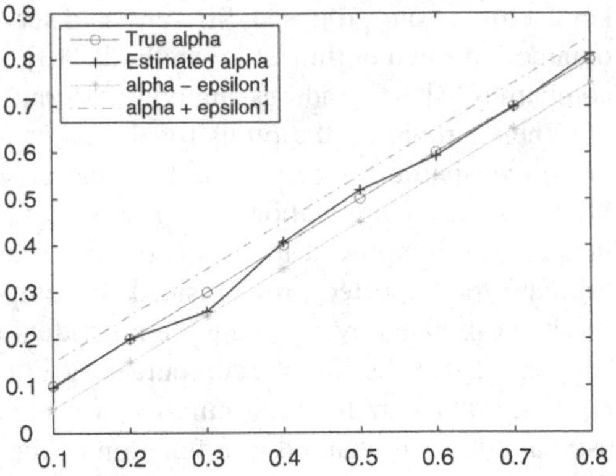

Fig. 5.3 Estimated α values
vs. simulated true α values

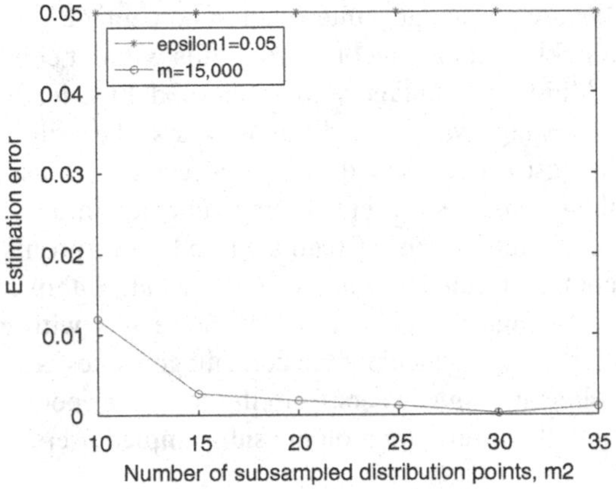

Fig. 5.4 Estimation errors
$|\hat{\alpha} - \alpha|$ vs. sub-sampling
number m from the entire
distribution

   To show the impact of different values of m as the sub-sampled number of
data points, we use the data set generated with $N = 100,000$ and $\alpha = 0.7$. The
parameters $\epsilon = 0.05$, $\delta = 0.05$ are fixed. We compute the estimation error with
varying m values. The result is shown in Fig. 5.4. As can be seen, with a larger
sub-sample number m, the estimation error generally becomes smaller, which is
consistent with the spirit of sublinear algorithms: the more we sub-sample, the more
precise will be the results that we obtain. However, when the result becomes closer
to the true value, we need much more sub-samples, i.e., if we want to further reduce
the error that is already small, we need to give a much greater increment for m.

5.6 Summary and Discussions

In this chapter, we presented an application of sublinear algorithms to load profile
classifications and differentiated services based on real smart meter data of a large
scale. The core part of our approach was finding an existing sublinear algorithm

that was suitable for our classification objective, yet could not be directly applied because of a gap in the confidence estimation. We thus proposed a sublinear algorithm that calls the existing sublinear algorithm a sub-function and completes the confidence estimates for our problem. The proposed algorithm inherits the original sublinear algorithm and makes no assumption on the structure of the distributions, which makes it robust. A simulated data set was used to evaluate our proposed methods and validate our approaches as efficient and accurate in a well-bounded estimation.

References

1. S. Amin and B. Wollenberg, "Toward a smart grid: power delivery for the 21st century", in *IEEE Power and Energy Magazine*, vol. 3, no. 5, pp. 34–41, 2005.
2. T. Batu, L. Fortnow, R. Rubinfeld, W. Smith, and P. White, "Testing that distributions are close", in *Proc. IEEE FOCS'00*, Redondo Beach, CA, Nov. 2000.
3. S. Bu, R. Yu, and P. Liu, "Dynamic pricing for demand-side management in the smart grid", in *Proc. IEEE Online Conference on Green Communications (GreenCom)*, New York, NY, Sept. 2011.
4. S. Chen, K. Xu, Z. Li, F. Yin, and H. Wang, " A privacy-aware communication scheme in Advanced Metering Infrastructure (AMI) systems", in *Proc. IEEE Wireless Communications and Networking Conference (WCNC)*, pp. 1860–1863, Shanghai, China Apr. 2013.
5. A. Czumaj, F. Ergun, L. Fortnow, A. Magen, I. Newman, R. Rubinfeld, and C. Sohler, "Sublinear-time approximation of Euclidean minimum spanning tree", in *Proc. SODA'03*, Jan. 2003.
6. F. Ergun, H. Jowhari, and M. Saglam, "Periodicity in Streams", in *Proc. Random'10*, Barcelona, Spain, Apr. 2010.
7. W. Fan and A. Bifet, "Mining Big Data: Current Status, and Forecast to the Future", in *SIGKDD Explor. Newsl.*, vol. 14, no. 2, pp. 1–5, 2013.
8. H. Farhangi, "The path of the smart grid", in *IEEE Power and Energy Magazine*, vol. 8, no. 1, pp. 18–28, 2010.
9. V. Ford and A. Siraj, "Clustering of smart meter data for disaggregation", in *Proc. IEEE Global Conference on Signal and Information Processing (GlobalSIP)*, Austin, TX, Dec. 2013.
10. Z. Ghahramani and M. Jordan, "Supervised learning from incomplete data via an EM approach", in *Advances in Neural Information Processing Systems 6*, San Francisco, CA, 1994.
11. V. Gungor, D. Sahin, T. Kocak, S. Ergut, C. Buccella, C. Cecati, and G. Hancke, "Smart Grid Technologies: Communication Technologies and Standards", in *IEEE Transactions on Industrial Informatics*, vol. 7, no. 4, pp. 529–539, 2011.
12. Roderick J. A. Little, "Regression with missing X's: A review", in *Journal of the American Statistical Association*, vol. 87, no. 420, pp. 1227–1237, 1992.
13. L. Qian, Y. Zhang, J. Huang, and Y. Wu, "Demand Response Management via Real-Time Electricity Price Control in Smart Grids", in *IEEE Journal on Selected Areas in Communications*, vol. 31, no. 7, pp. 1268–1280, 2013.
14. R. Rubinfeld and A. Shapira, "Sublinear Time Algorithms", *SIAM Journal on Discrete Mathematics*, vol. 25, no. 4, pp. 1562–1588, 2011.
15. M. Roozbehani, M. Dahleh, and S. Mitter, "Dynamic Pricing and Stabilization of Supply and Demand in Modern Electric Power Grids", in *Proc. IEEE Smart Grid Communications (SmartGridComm)*, Gaithersburg, MD, Oct. 2010.

16. S. Shao, T. Zhang, M. Pipattanasomporn, and S. Rahman, "Impact of TOU rates on distribution load shapes in a smart grid with PHEV penetration", in *IEEE Transmission and Distribution Conference and Exposition* New Orleans, LA, Apr. 2010.
17. D. Wang, Y. Long, and F. Ergun, "A layered architecture for delay sensitive sensor networks", in *Proc. IEEE SECON'05*, Santa Clara, CA, 2005.
18. Q. Wang, M. Liu, and R. Jain, "Dynamic pricing of power in Smart-Grid networks", in *Proc. IEEE Decision and Control (CDC)*, Maui, HI, Dec. 2012.
19. C. Wong, S. Sen, S. Ha, and M. Chiang, "Optimized Day-Ahead Pricing for Smart Grids with Device-Specific Scheduling Flexibility", in *IEEE Journal on Selected Areas in Communications*, vol. 30, no. 6, pp. 1075–1085, 2012.

Chapter 6
Concluding Remarks

6.1 Summary of the Book

This book is divided into two parts: the foundation part (Chap. 2) and the application part (Chaps. 3–5). In Sect. 2.3.1, we start from an easy application of inequalities to derive a very first bound. We then study finding distinct elements in Sect. 2.3.2. We show an insight to solve the problem and how to analyze the insight. In Sect. 2.3.3, we present a two cat problem and we develop an algorithm that is sublinear, yet differs from the traditional $(1 + \epsilon, \delta)$ sublinear algorithm format. We then look into three applications of sublinear algorithms on wireless sensor networks, big data processing, and smart grids in Chaps. 3–5.

In Chap. 3, we look at an application on wireless sensor networks. It shows how to go from simple properties development into an application design. To collect data in wireless sensor networks, one may consider developing an indexing structure, so that the sensors are organized and a query can go to individually selected sensors. However, such organization is less scalable. The layered architecture is superior in that it is extremely simple and purely distributed. The query results have quality guarantees. These are the benefit of sublinear algorithms. This layer architecture is suitable to serve as a first round check so that an in-depth investigation can be carried out much less frequently.

In Chap. 4, we look at a data skew problem in the MapReduce framework. There are many studies addressing such data skew problem. For example, there are reactive approaches, which try to monitor the data skews and move data from heavy-loaded machines to light-loaded machines. We instead study proactive online algorithms. Our first algorithm is a straightforward greedy algorithm. Our second algorithm is developed under an observation where we first find heavy keys and then assign the data to respective machines by balancing the heavy keys. We demonstrate that by better allocating heavy keys, we can achieve better performance. To find the heavy

© The Author(s) 2015
D. Wang, Z. Han, *Sublinear Algorithms for Big Data Applications*,
SpringerBriefs in Computer Science, DOI 10.1007/978-3-319-20448-2_6

keys, we use a sampling technique. We analyze the amount of samples needed. In practice, a proactive sublinear algorithm can work together with a reactive approach to further improve the performance.

In Chap. 5, we look at a user classification problem in smart grid. The classification is based on user behavior in their electricity usage. The potential usage of such classification is to develop differentiated pricing services for different type of users. We first present a trace study using real world data collected by smart meters. We show that user behaviors are indeed different. We conduct classification based on user electricity usage distributions. As compared to peak electricity usage, average electricity usage, the electricity usage distribution maintains more information of a user. The classification face a big data problem as there easily have millions of users, and for each user, his electricity usage data are big. Our treatment of this problem is to borrow an existing theoretical study on testing whether the two distributions are close. The algorithm is then revised according to the smart grid application scenario.

6.2   Opportunities and Challenges

In the past, the computing community focuses on computing intensive applications. A good example for computing intensive applications is playing chess. The amount of input data is minimal; yet the computational complexity is huge. Nowadays, we are facing an increasing amount of data. To achieve a certain task, the data we need to process increase from gigabytes to terabytes and to petabytes. In data intensive applications, the computing process of each piece of data can be trivial. The processing time for current applications with a big data flavor is dominated not only by computing but more by I/O access of the data. To make things worse, because the data is big, the data often have to be stored in hard disk. This makes the I/O access a disk access rather than memory access and the access time substantially increased.

Consequently, if an algorithm that works less than linear time is only of theoretical importance, and is a fantasy in the past, it becomes a necessity today. We see immense applications scenarios and thus opportunities.

Sublinear algorithms are usually very simple to implement and are distributed in nature. The guarantee bound is valuable if certain service level agreement is required by the application. In the applications where frequent initial checking is necessary before an in-depth analysis can be carried out, a guarantee in the initial results is also valuable. These are the domains where sublinear algorithms help most.

There are many challenges. Sublinear algorithms heavily bear the characteristics of algorithms. To develop algorithms, one needs insights and analysis of the insights. This means that in many occasions, developing a sublinear algorithm and its analysis can be a case-by-case art.

Currently, there are more studies on developing sublinear algorithms from the theoretical computing science point of view. There are many surveys, tutorials and books with nice collections of different sublinear algorithms. To date, there are studies ranging from checking distinct elements, Quantile, heavy hitters, to connectivity, min-cut, bipartiteness, minimum spanning trees, average degree and to whether a data stream is periodic, whether two distributions are close. Similar to what we have learned in finding distinct elements in Chap. 2.3.2, these algorithms have interesting insights and analysis.

We argue that learning more sublinear algorithms, their insights and analysis is essential in mastering sublinear algorithms. Nevertheless, this book specifically tries to avoid becoming a collection of sublinear algorithms. This book tries to explain some nuances of using sublinear algorithms in applications.

Applying sublinear algorithms to application scenarios, one may fit existing sublinear algorithm results to specific application scenarios. However, applying existing results to an application scenario does not have a step-by-step procedure. The research in applying sublinear algorithms to real world applications is still in its infancy and more work needs to be done.

In addition, it is often the case that a real world application will not be solved by a single sublinear algorithm. A joint force of sublinear algorithms and other techniques is necessary. Formulating a problem and addressing it by partially using sublinear algorithms are challenging.

Printed in the United States
By Bookmasters